PSYCHOLOGY OF FRAUD

Integrating Criminological Theory into Counter Fraud Efforts

J. MICHAEL SKIBA, PHD,
"DR. FRAUD™"

PSYCHOLOGY OF FRAUD
INTEGRATING CRIMINOLOGICAL THEORY INTO COUNTER FRAUD EFFORTS

Copyright © 2017 J. Michael Skiba, PhD, "Dr. Fraud™".

All rights reserved. No part of this book may be used or reproduced by any means, graphic, electronic, or mechanical, including photocopying, recording, taping or by any information storage retrieval system without the written permission of the author except in the case of brief quotations embodied in critical articles and reviews.

iUniverse
1663 Liberty Drive
Bloomington, IN 47403
www.iuniverse.com
1-800-Authors (1-800-288-4677)

Because of the dynamic nature of the Internet, any web addresses or links contained in this book may have changed since publication and may no longer be valid. The views expressed in this work are solely those of the author and do not necessarily reflect the views of the publisher, and the publisher hereby disclaims any responsibility for them.

Any people depicted in stock imagery provided by Thinkstock are models, and such images are being used for illustrative purposes only.
Certain stock imagery © Thinkstock.

ISBN: 978-1-5320-2218-0 (sc)
ISBN: 978-1-5320-2219-7 (hc)
ISBN: 978-1-5320-2217-3 (e)

Library of Congress Control Number: 2017908428

iUniverse rev. date: 10/13/2017

To my wife, Karen, and my children, Gabrielle and Julia, who inspire me every day

Contents

Acknowledgments xi
About the Author xiii
Introduction xv

CHAPTER 1: FRAUD: WHAT'S THE PROBLEM? 1
    Damaging Effects of Fraud 4
    Insurance Fraud Is Difficult to Detect 6
    Public Acceptance of Insurance Fraud 8
    Prosecutorial Efforts 10
    Legal Concept of Bad Faith 12

CHAPTER 2: CRIMINOLOGICAL THEORY 15
    Schools of Criminology 17
    Biological School of Criminology 18
        Genetic and congenital foundations of antisocial behavior 21
        Genetic factors in victimization 21
        Brain dysfunction and criminal behavior 22
    Psychological School of Criminology 23
    Sociological School of Criminology 25
        Routine activities theory 26
        Strain theory 27
        Deterrence theory 29

    Rational choice theory 29
  Fraud Application of Theories 32
  Biological School of Criminology and Fraud Application 32
  Psychological School of Criminology and Fraud Application 34
  Sociological School of Criminology and Fraud Application 35
    Routine activities theory and insurance application 36
    Strain theory and insurance application 42
    Deterrence theory and insurance application 44
    Rational choice theory and insurance application 45

CHAPTER 3: THE FRAUD TRIANGLE 49
  Pressure 51
  Rationalization 55
  Opportunity 60

CHAPTER 4: COUNTER FRAUD EFFORTS 64
  Identifying Risks/Vulnerabilities 65
  Assessing Vulnerabilities 77
  Completing the Vulnerability Assessment Document 85
  Developing Red Flags 97
  Integrating Controls 105
  Monitoring and Modifying Controls 116
  Analyzing Measurement Criteria 121

CHAPTER 5: DEVELOPING A MULTIPRONGED APPROACH TO COUNTER FRAUD 123
  Psychological Profile of a White-Collar Criminal 124
  Honesty and Cheating Behavior 130

Checks and Balances 143
Vacations? 145
Code of Conduct 145
Culture of Honesty 154
Hiring Practices 156
Internal Training 159
Wages and Internal Theft—Entitlement Theory 161
Hotlines 162
Information Sharing 164
Legislation 166

Chapter 6: The Behavioral Bridge 170
Behavioral Bridge: Defined 170
Behavioral Pattern Predictors 172
Behavioral Bridge Applications 173

Chapter 7: Final Thoughts 178
Fraud Is a Broader Social Problem 179
Application of Vulnerability Theory 181
The Deterrent Effect 184
Measurement of Fraud 186
    Line of Business 188
    Foundation 189
    Formula 191
    Fraud Measurement Recommendation 195
International Perspective 197
    Canada 199
    Malaysia 200

 Italy 202
 Germany 203
 Switzerland 206

Conclusion 209
References 213
Index 223

Acknowledgments

I wish to acknowledge the many people who made this book possible. First, thank you to all my professional counter fraud colleagues who provided insights at various stages of the book; I am honored to serve with such professional and outstanding individuals. My hope is that this final product will serve the anti-fraud community well, adding to the knowledge base in this unique criminal area. Second, to all my academic colleagues, who helped me navigate when I reached those brick wall moments. Third, to my family and friends for their patience and encouragement. Finally, to my wife, Karen, who helped me keep my eye on the prize and reach the finish line, and to my children, Gabrielle and Julia, who truly amaze me every single day. Thank you all for your support through my journey.

About the Author

J. Michael Skiba, "Dr. Fraud,™ " is an international expert on economic crime and has been referred to as one of the top crime fighters in the world. He holds an MBA and a PhD with a research focus on fraud and economic crime and is considered one of the leading researchers in this area. He has been a college professor for more than ten years and currently holds the role of Program Chair of Criminal Justice at Colorado State University- Global Campus. In this leadership role, he oversees the university's bachelors- and masters-level criminal justice department, including specializations in fraud management, criminology, and forensics. He is also asked to lecture on economic crime at universities around the world, including recent sessions in Germany, Switzerland, and the Netherlands.

Dr. Skiba's background in white-collar crime spans a twenty-year time frame, which started at Allstate Insurance, continued to Interboro Insurance, and led him to his current role as an international consultant. He is currently consulting with INFORM, an international cybercrime company, and holds the position of Vice President of International Counter Fraud Strategies, where he oversees and manages the development of global counter fraud programs. During his career at Allstate and Interboro, he held many positions, including Senior Claims Consultant, Special Investigations Field Investigator, Analyst, and Major Case Coordinator. He was also extremely active in the industry and was Vice President of the International Association of Special Investigative Units, President of the New York State Chapter of Special Investigative Units, member of the Insurance Emergency Operations Command, and board member of the New York State Department of Criminal Justice Services-Motor Vehicle Theft and Insurance Fraud Board.

As a consultant for more than thirteen years, he has assisted dozens of companies and agencies develop extremely successful counter fraud programs. He is a highly sought-after speaker on the topic of insurance fraud and regularly provides training at conferences and to corporations and agencies, including recent engagements throughout the United States, Austria, Vancouver, Sweden, London, and Brazil. He was also the keynote commencement speaker at the 2016 graduation ceremony for Colorado State University- Global Campus. The Bureau of Alcohol, Tobacco, Firearms, and Explosives (ATF) regularly calls upon him to train special agents and prosecutors on fraud and economic crime. In addition, he has performed numerous TV and radio appearances, where he is asked for his commentary on economic crime issues. Dr. Skiba provides a unique global perspective on economic crime, one that serves as an extremely strong foundation for this book.

Introduction

Insurance fraud is a crime that is continuing to grow in frequency and intensity at a very alarming rate. Counter fraud professionals working in this arena will testify to the incredible damaging impact this crime has, yet it seemingly does not get the resources and attention that it deserves in order for these professionals to adequately develop strong counter fraud strategies. Street-smart criminals are aware of this vulnerability and undoubtedly take advantage of these weak controls. The endeavor of this book is to outline key preventative strategies that can be implemented in any fraud agency or organization—strategies that will assist with developing highly focused policies and programs that help to combat this crime. From a crime-fighting perspective, these preventative efforts will force fraudsters to rethink their criminal activity and consider other easier targets and criminal areas. From a business perspective, these focused strategies will serve to reduce internal costs, thus positively affecting a company's combined ratio and increase one's competitive position in a saturated insurance market. No matter what the perspective, it is apparent that fighting fraud should be a top priority.

In the first chapter of the book, we will delve into the fraud problem and develop a deep understanding of how damaging this specific crime is. We will look at fraud trends and patterns and focus on why this crime is becoming the *preferred* crime for all fraudsters in the spectrum, from organized criminal groups, terrorist organizations, and even smaller-scale opportunistic fraudsters. We will learn that fraud not only causes financial damage but has societal and humanitarian impact as well.

Fraud is extremely difficult to quantify; thus one of the most significant challenges lies in simply understanding the nature and scope of this mounting issue.

In Chapter 2, we will explore criminological theory, an exciting integration of psychology and crime, and learn how these principles can be applied in a counter fraud setting. Criminological theory helps us to understand the mind of a criminal, further delving into the cognitive aspects of what motivates offenders; yes, we will explore the mind of a fraudster. We will begin by presenting the foundational concepts of these criminological theories and then discuss how they have been successfully applied in other criminal areas yet remain to be utilized in a counter fraud setting. This is one of the most significant areas of exploration for this book, the application of criminal theory into anti-fraud efforts. We will see how difficult it is to develop effective counter fraud efforts, as current policies are mainly founded on unreliable, unsupported, unacademic data. It is the endeavor of this book to present a contemporary perspective on integrating theory into anti-fraud programs. The focus of these contemporary strategies is on reducing the opportunity to commit fraud by addressing one's vulnerabilities and then developing solid counter fraud strategies based on these specific weak areas. We know that companies cannot focus on everything, so the recommendation is to focus on those areas that are the most important and have the most damaging impact.

In Chapter 3, we will analyze the fraud triangle and investigate how it has been, and should be, applied in an insurance fraud setting. We will expand on its three key elements—pressure, opportunity, and rationalization—and explore each in depth in order to comprehend the value of the triangle in counter fraud strategies. Pressure stems from the push factor and could be financial debt or a recent life-changing event such as divorce, a new child, job loss, or medical condition. Opportunity is seen as one area of the triangle that can be manipulated and controlled by counter fraud professionals. We should consider programs that help reduce opportunity and increase the deterrent effect. Rationalization focuses on how offenders justify committing fraud; this could stem from various factors, such as trying to recoup premiums or a sense of entitlement.

Chapter 4 will focus on specific counter fraud efforts and explore various methods for application, including the self-designed vulnerability assessment, a focused procedure for identifying risks and then appropriate fraud strategies. A full review of this procedure will be provided in order for readers to utilize this protocol within their respective agency or company. This assessment is a formal method to assess the strengths and weaknesses of current fraud systems and then develop more effective, focused strategies. This formalized approach has been field-tested and utilized in more than a dozen organizations and has been deemed highly effective in creating these focused strategies. In the remaining sections of chapter 4, the detailed steps to developing counter fraud efforts will be presented, which include identifying risks and vulnerabilities, developing red flags, integrating controls, and monitoring and modifying those controls. These are all critical steps in the counter fraud process, and, accordingly, each will be discussed in detail. The vulnerability assessment will assist with the first step of identifying risks; developing strong red flags as a result of the assessment will be the next step, flags that will serve as a foundation for a highly focused fraud strategy. A very critical piece of this process is integrating controls; this is where the rubber meets the road. It must be decided what system of controls will be utilized: internal, external, vendor supported, or internally supported. Furthermore, decisions need to be made as to how these systems will be measured and then adjusted accordingly. A final step is monitoring and modification, which is often a neglected part of this process, yet it is extremely important for the success of an anti-fraud program.

In Chapter 5, we will examine the benefits of developing a multipronged approach to fraud prevention—that is, developing a blueprint that will serve to thwart fraud efforts from many different perspectives. Fraudsters are motivated by different factors, and thus different approaches are needed for effective preventative efforts. We will examine the profile of a white-collar criminal, honesty and cheating behavior, checks and balances, vacations, codes of conduct, corporate culture, hiring practices, internal training, entitlement theory, hotlines, information sharing, and legislation and how they all can be used as part of a global fraud campaign.

In Chapter 6, I am very excited to unveil a self-developed approach to prevention: the behavioral bridge. This approach focuses on behavioral pattern predictors (BPPs)—that is, behavior that can be measured in data format and then inserted into an analytical system for highly detailed, highly effective structured preventative efforts. In simple form, it is the bridge between big data and specific preventative efforts; I think you will find this approach very applicable and a highly effective technique for application.

In Chapter 7, the last section of the book, we will present final thoughts and explore various, more global, aspects that should be considered as part of successful counter fraud strategy. We will discuss public awareness and how fraud is seen as a social problem in need of global education. Fraud is seen as a victimless misdeed that has evolved into a socially acceptable crime, which undoubtedly contributes to the negative cycle of fraud that ensues. We will then discuss the measurement of fraud and how, as an industry, we do not have a clear understanding of the problem, which leads to improper decisions, a lack of comprehension of where to focus efforts, and misinformation regarding the true impact of this crime. I will introduce a self-developed three-step proposal toward the measurement of fraud, presenting ideas to foster a consistent method for more accurate fraud computation. In addition, recommendations will be made as to a specific formula for fraud measurement that can be applied immediately. International issues will also be a focal point of this last chapter; we will examine some of the similarities and differences in fraud prevention in different geographical areas and look to determine the efficacy of their approaches.

We have a lot to cover, so let's dive right in!

CHAPTER 1

Fraud: What's the Problem?

Estimates reveal that insurance fraud costs an incredible $80 billion per year in the United States, which translates into almost $1,000 per household annually (Coalition Against Insurance Fraud, n.d.). Fraud, a category of white-collar crime, has been declared by the United States Department of Justice as the number two crime facing the United States. Weak preventative controls and low punishment make this unique crime attractive to the criminal element, so much so that it is becoming the preferred method of financial funding for many organized criminal and terrorist groups. Fraud is seen as a low-risk, high-reward offense; thus, when compared to other types of crimes, such as extortion and drugs, fraud has become very appealing to the criminal element. From a business perspective, insurance professionals estimate that fraud results in approximately 10 percent losses each year, leading to a very disturbing financial impact on the insurance industry.

PSYCHOLOGY OF FRAUD

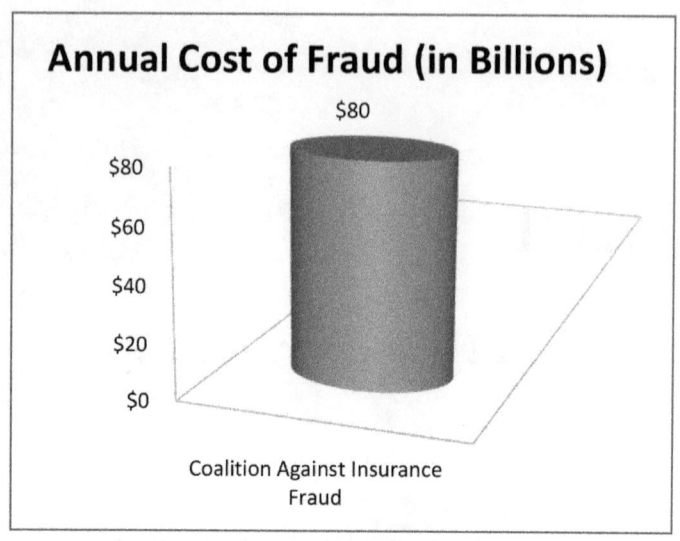

It is commonly known in the fraud industry that approximately 10 percent of all claims are suspicious, which is an incredibly significant percentage. However, in my opinion, based on performing dozens of audits and also reviewing industry studies, this percentage is quite higher. One such closed claim study found that 21 percent of bodily injury claims and 18 percent of personal injury protection claims had some degree of fraud or buildup, an extremely troublesome figure (Insurance Research Council 2015). It is established that fraud is a true problem worthy of further analysis, research, and exploration. But what exactly is fraud, and how is it defined? The following definitions are a compiled summation of fraud terms taken from seminal sources, such as the National Insurance Crime Bureau, Insurance Research Council, and the Coalition Against Insurance Fraud. These definitions will serve as a reference point for this book.

Insurance fraud. An act that a person or entity, individually or jointly, willfully commits to obtain monetary gain from an insurer by knowingly presenting false evidence of economic loss.

Hard fraud. The fraud offender commits the actual act that leads to the loss, such as arson, a staged accident, or home robbery. The actions of the fraudster are prosecutable and often involve larger sums of money.

In these cases, the fraud is illegal, but quite often so is the actual act, thus making this criminally prosecutable.

Soft fraud. The most widely committed type of insurance fraud, and often involves the misrepresentation of facts. This scenario involves an individual who suffers a legitimate loss but inflates and exaggerates the claim in order to gain a higher payout.

No-Fault fraud. No-fault fraud is any fraud that involves taking advantage of the medical structure of states that carry no-fault laws. This fraud typically involves medical provider fraud—that is, medical facilities that overbill, or bill for services not rendered, in order to increase reimbursement.

Padding. A form of soft fraud involving the overstatement of the actual value of a claim for the purposes of increased financial gain by the policyholder. Padding often occurs in automobile cases whereby an insured attempts to inflate damages to recoup the deductible.

Planned fraud. This term is used synonymously with *hard fraud* and refers to instances whereby an individual or group fabricates certain elements leading up to a loss; this could include filing a false claim or staging an entire fictitious incident. The main distinction with planned fraud is that it involves a cognitive, contrived effort to create a fraudulent scenario. An example would be a spouse who loses her wedding ring on vacation and files a claim stating that it was stolen from their residence.

Opportunistic fraud. This term is used synonymously with *soft fraud* and occurs when an individual tries to claim additional damage after a legitimate loss occurs. These claims can be more difficult to detect, as they are legitimate claims with few traditional red flag indicators. These individuals are often upstanding citizens that have no history of prior fraudulent activity, yet are presented with an opportunity to inflate and exaggerate and capitalize on the opportunity. An example would be the same spouse who suffered a legitimate theft of her wedding ring from her residence and also decides to claim that other pieces of jewelry were stolen as well.

Medical provider. A hospital, medical group, treatment center, doctor, or any other facility that provides medical services to patients.

SIU (Special Investigative Unit). An investigative unit within a private sector insurance carrier or company or law enforcement agency that focuses primarily on investigating fraud.

White-collar crime. Those illegal acts that are characterized by deceit, concealment, or violation of trust, and which are not dependent upon the application or threat of physical force or violence. Individuals and organizations commit these acts to obtain money, property, or services; to avoid the payment or loss of money or services; or to secure personal or business advantage.

Damaging Effects of Fraud

Fraud has significant financial impact, costing us billions of dollars per year, but how exactly is this figure calculated? The calculation of fraud's financial impact includes money directly stolen from innocent fraud victims by unscrupulous contractors and other service providers and also includes the increased premiums that result as insurers pass along the high costs of fraud to policyholders. Businesses must pass along the costs of higher premiums to their customers by increasing the costs of their consumer goods or services in order to attempt to recoup some of this capital. These businesses also raise consumer costs as they try to offset the high price of health and other types of business insurance.

There are also other associated damaging effects of fraud. One unfortunate consequence of insurance fraud is its negative impact on legitimate patient care; fraudulent medical billing practices have been shown to result in significant abuse to patients. The research has revealed that questionable medical facilities will inflate billing (up-coding) in order to increase the reimbursement that they receive from the insurance carriers. This up-coding procedure often results in patients being subject to highly painful procedures, such as invasive testing, unnecessary surgeries, and extreme physical manipulations, which can result in severe injury, including several documented cases of paraplegia and death. A common scheme with fraud rings is to subject patients to an extensive electromyography (EMG) test, which is very uncomfortable and painful, just for the mere purpose of inflating the medical costs. Other rings have been known to use children as part of a staged accident scenario in order to add legitimacy to the loss; the rationale from the ring leader is that if children are included in the loss, the insurance carrier would assume it was legitimate and not staged. These children are subjected to painful procedures, with the only goal of bill inflation. Residential and commercial arson cases have also resulted in severe injury and death. There have also been documented cases of fraudulent life insurance schemes involving murder in order to collect high payouts.

One of the most significant large-scale cases of dangerous and unnecessary treatment occurred at Redding Medical Center; a federal investigation revealed that over a three-year period, the facility performed unnecessary cardiac catheterizations and bypass surgeries on approximately 769 patients; these procedures included life-threatening open-heart and bypass surgeries, all for the sole purpose of increasing billable amounts. These unnecessary procedures created additional medical problems such as stroke, heart attack, and paralysis to patients who underwent them (Charatan 2002).

Innocent patients are also placed at increased risks from overzealous doctors through prescriptions, testing, therapy, and surgical treatment. There are many documented cases of physicians who allow medical procedures to be performed by unlicensed and untrained personnel, placing patients at incredible health risk. Additional consequences can be seen in the increased cost of health care that results from fraud. Fraud schemes drain so much from our health care system that it leaves a large majority of people uninsured due to high premium costs, which correlates to reduced care as these individuals cannot afford adequate health coverage for themselves or their families.

Reduced patient care is also a frequent issue in the pharmaceutical industry. Unscrupulous pharmaceutical companies use aggressive and deceptive marketing practices, promote drugs not approved by the Food and Drug Administration, price manipulate, and use improper detailing procedures in order to increase profits. Prescription drug spending totaled $457 billion in 2015 (Office of the Assistant Secretary for Planning and Evaluation 2016), and continues to rise significantly each year; thus, there is incredible monetary potential to entice unscrupulous activity. Drugwatch.com is a very useful site, which lists common prescriptions that were sold without being approved by the FDA. For example, Pfizer pharmaceutical company paid $2.3 billion in a settlement for marketing drugs that were not approved by the FDA. The consumers that purchased the product experienced severe health complications. Feel free to visit the following link to listen to the FBI audio file that summarizes the investigation: https://www.fbi.gov/audio-repository/news-podcasts-inside-pfizer-2.3-billion-settlement.mp3/view.

Insurance Fraud Is Difficult to Detect

Despite alarming facts about this mounting criminal issue, little is known about this crime; there are very few sources of concrete information about the insurance fraud problem and, furthermore, the most effective methods to combat this unique crime. In the academic world, fraud is extremely understudied, which presents additional challenges as there are few academic, peer-reviewed articles for counter fraud professionals to consider in their fraud strategies. During the course of my PhD journey, my dissertation committee was extremely excited that I was researching fraud, as it is a topic that is in its infancy in the academic world. Many researchers argue that fraud is a relatively invisible crime, one that is difficult to detect and quantify, making it one of the most worthy areas for further study. Many states require mandatory fraud reporting, but there is often a lack of communication beyond the state level; thus, each company or state and federal agency must devise its own reporting protocols, which are often inconsistent and inconclusive. The lack of accurate reporting is very typical in the white-collar crime arena; there is a large void in data collection and studies performed in the broader area of white-collar crime, to the extent that it hinders the knowledge base and causes increased challenges for those professionals attempting to fight this crime. Without an accurate measure of the impact of this crime, it is difficult to gain buy-in from claims and insurance executives when the time comes to allocate resources to fraud units. This brings an incredible sense of frustration to those working in fraud units, who are aware that this is a huge problem yet cannot adequately quantify and present data to those in management positions.

An additional challenge, as confirmed by the research, shows that there is no universal definition of fraud, which leads to inconsistent data on the fraud problem and makes it a difficult crime problem to quantify. While conducting training sessions, I will often begin a seminar by asking the attendees for basic definitions of fraud; the results from the audience are always diverse and inconsistent. To further this point, I performed an informal study during one audit and asked six auditors to review the same claim files from a fraud perspective. Each of the reviewers identified 5 to 10 percent of the claims as suspicious (which is consistent with industry standards), yet there was not one specific claim that was deemed fraudulent by all six auditors. This smaller-scale example illustrates the true nature of this crime; it is difficult to quantify!

Other than research that I have personally performed as part of my dissertation or other projects, the only foundational study ever performed on the measurement of fraud that I am aware of was completed in 2003 by the Coalition Against Insurance Fraud (see reference page for pdf). The key findings were as follows:

- Eighty-six percent of the participants track the percentage of claims referred to SIU.
- Thirty-nine percent (the majority) had an average referral rate of 1 to 3 percent; of significance was that 25 percent reported this rate was less than 1 percent.
- Fifty-seven percent (the majority) would find it extremely useful to compare their rate to an industry average.

- Eighty percent track anti-fraud savings.
- Twenty-six percent (majority) use the claim reserve to calculate savings, 24 percent use the actual payout, 22 percent use compromises, 18 percent use the total claims payout, and the remaining 10 percent use other methods.
- Ninety percent felt that their calculated fraud savings is an inaccurate reflection on the real savings of SIU investigations.

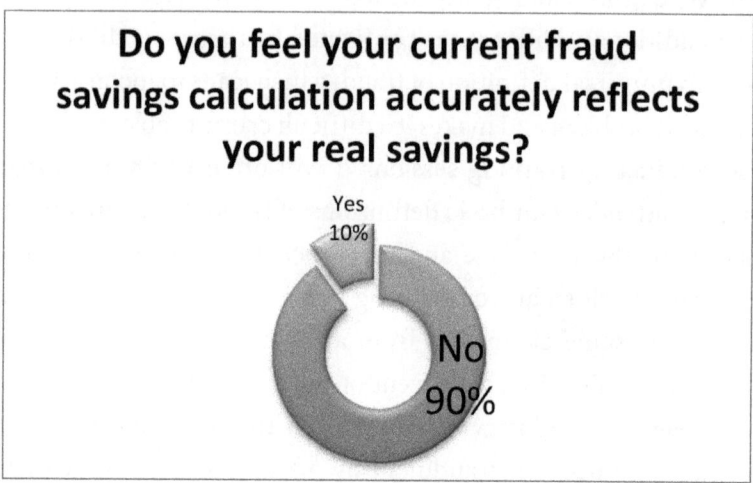

- Ninety-seven percent felt that their calculated fraud savings as reported to state fraud bureaus does not accurately reflect anti-fraud efforts.

Public Acceptance of Insurance Fraud

Despite insurance fraud's documented, broad-reaching impact, research has revealed alarming statistics in regard to the public's opinion toward this crime. Approximately 24 percent of the population believes that it is acceptable to exaggerate the value of an insurance claim, and 11 percent believe that it is acceptable to submit a claim for items or damages not actually lost.

Furthermore, 30 percent agree that fraudulent activity will increase during an economic downturn, and 49 percent felt that they would not be caught if they filed a fraudulent claim. Eighteen percent felt it is acceptable to pad a claim to recover premiums and 24 percent to pad a claim to recover a deductible (Coalition Against Insurance Fraud, n.d.; Insurance Research Council 2013; Ishida, Chang, and Taylor 2016; Skiba and Disch 2014).

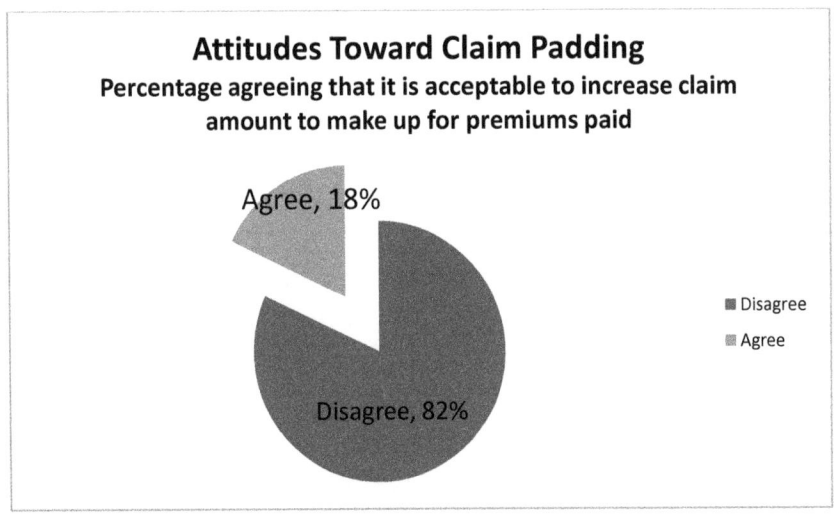

Many researchers postulate that public acceptance is due to the negative perception of insurance companies and how the publication of their high profits helps to justify consumer fraudulent activity. For this reason, there seems to be a lack of social pressure to enact, enforce, investigate, and prosecute insurance fraud-related issues.

Research has attempted to explain the root cause of this overwhelming acceptance of fraud. Historically, insurance fraud was often ignored because its costs could simply be passed on to consumers through increased premiums; there seemed to be no identifiable victim of the crime. In later chapters, we will explore how this *victimless* perspective contributes to the overall fraud problem. Interestingly, this lack of historical enforcement may have led to a generalized paradigm of social acceptance; in other words, the acceptance of this crime has slowly evolved. It is further argued that consumers who feel that insurance companies make too much money, treat policyholders unfairly, and charge excessive premiums, will be more tolerant of insurance fraud. This creates a very damaging cycle of fraud, a cycle that needs contrived intervention.

Prosecutorial Efforts

Fraud has been an issue for many years, yet it still has not received the prosecutorial support that it deserves. Inconsistent court rulings, small-scale penalties, and unclear criminal statutes contribute to the lack of prosecutorial support. Looking at this from a historical and global perspective, it is important to note that the post 9/11 era has drawn resources from white-collar crime prevention toward terrorist efforts. Studies have revealed that one of the significant contributors of this lack of prosecutorial attention is the inadequate funding and resources to support insurance fraud strategies, citing how these cases are very expensive to prosecute. Post-9/11, many law enforcement and judicial agencies transferred resources from white-collar crime units to terrorism-related units; yet research revealed that fraud continued to rise in scope and breadth during that same time period, highlighting that it is a neglected area of judicial enforcement.

When compared to other types of crimes, insurance fraud and other forms of white-collar crime are very expensive to prosecute, largely in part due to their complex structure, which involves multiple agencies and jurisdictions. Research has revealed that most fraud cases that are successfully prosecuted are done so by prosecutors funded from areas other than strict state revenue. Alternative prosecutorial funding should be considered so that the insurance industry can capitalize on these opportunities to increase the prosecution of fraud cases (Abramovsky 2008). Several companies I have consulted with have seen very positive results in the use of state agency budgeting structure and how this can be useful to creatively allocate prosecutorial resources. However, participants should proceed with caution, as legal challenges may arise out of this structure focusing on how this funding can be perceived as creating favoritism from prosecutors toward the insurance companies that fund their offices.

Three models of insurance and governmental coordination have been developed. The first model was established in Massachusetts, and as such is named the Massachusetts Model. This approach involves the development of a quasigovernment agency whereby the directors are appointed by the insurance industry. The potential impartialness of this model was legally challenged in *Commonwealth v. Ellis* (1999), whereby the defendants alleged bias between the insurance company, the State Fraud Bureau, and the prosecution. The court concluded that the defendant was not denied any constitutional right as a result of the relationship between the three parties. The second approach, named the Majority Model, is followed by most states and involves the development of a specialized fraud bureau that is embedded in an established government division, such as that state's department of insurance or other governmental agency. The last model, which is present in eight states, is characterized by a lack of an established, formal fraud bureau; most insurance companies provide investigators to work directly with public offices (Abramovsky 2008).

Many states legislatively mandate insurance company activity in relation to fraud investigations. Studies have revealed that states with certain mandated investigative laws are the most effective at combating insurance fraud related crimes. Significant fraud reduction occurs in states that mandate insurers to establish special investigative units (SIUs) within their companies. Pennsylvania assesses the insurance company contribution for fraud prevention strategies based on the amount of overall business the company conducts within the state, and then those proceeds are allocated in a general fund for disbursement to prosecutors for use in insurance fraud prosecutions. Nevada has a similar program where insurance companies are assessed based on the premiums they charge within the state. These funds are then allocated to the Office of Attorney General, which is the agency primarily responsible for insurance fraud investigations.

Legal Concept of Bad Faith

The concept of good faith requires insurance companies to conduct themselves within the guidelines of policy provisions as outlined in an insurance contract with a policyholder. This concept was developed to regulate the conduct of insurers and ensure policyholders are treated fairly. This presents an interesting conundrum during fraud investigations, as policyholders are often the focal point of these damaging cases. An official business contract is established between the policyholder and insurance company when an insurance policy is written; this policy creates a legal agreement between both parties, including provisions that both will act in the utmost good faith in handling and disclosing information pertaining to the legal contract. At its core, it relies on trust and honesty on behalf of both parties to the contract.

In many states, policyholders may file a bad faith suit accusing the insurance company of breaching the good faith provision of the original business contract. This is a tort action that not only allows recovery of the basic benefit but also additional damages above and beyond the policy limits. A bad faith suit is a potentially dangerous legal predicament for a company, as risk of financial and image damages can be quite high. If an insurance company fails to settle a legitimate claim, the company can be held liable for excess verdicts for bad faith, which is a very potentially dangerous and undesirable situation. Fraud investigators must find a balance between justified fraud investigation and entering into a bad faith legal scenario. Studies reveal that states that allow bad faith actions are characterized by insurers who report higher claim costs, indicating that companies operating within these states may have a higher proclivity to settle questionable fraudulent claims to avoid bad faith allegations (Asmat and Tennyson 2014). Bad faith states are also characterized by a lower degree of investigative intensity, as insurers are cautious of bad faith allegations. These alarming facts illustrate how bad faith allegations can have a negative impact on the prevention of insurance fraud offenses, as companies operating within these parameters will have a tendency to settle instead of investigate. Due to the increasing competition in the insurance industry, and the desire for the insurance companies to retain as many policyholders as possible, fraud investigations could become a low priority due to bad faith cautiousness.

This contention is supported when one analyzes the results of bad faith reform in West Virginia. In 2005, West Virginia passed S.B. 418. The Third Party Bad Faith Act eliminated the rights of third-party claimants to file lawsuits against an insurance company if they felt that they were treated unfairly, or in *bad faith*. West Virginia eliminated this legal right for claimants and instead provided them with an administrative procedure as an alternative, thus removing this process from the litigation environment. As one can view from the graph below, paid losses declined significantly starting in 2005, when S.B. 418 was adopted, and continued to fall until the measurement conclusion in 2010. Quantifiably, this reform reduced costs approximately $200 million from 2005 to 2010, an extremely significant result (Insurance Research Council 2011, 2014).

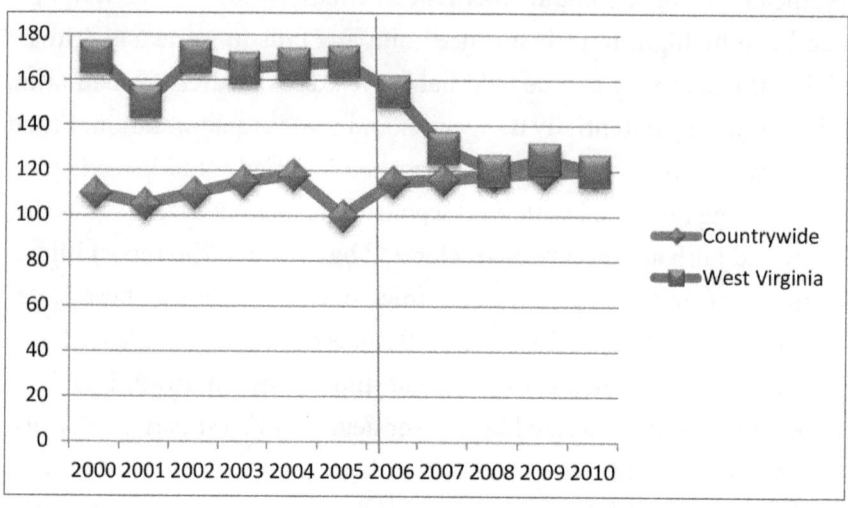

Bodily Injury Liability Coverage Loss Costs
(Paid losses per insured exposure)

In this chapter we came to understand that fraud is a definite problem worthy of further analysis and exploration. There are significant financial and humanitarian damages as a result of fraud, and companies must be able to focus efforts, or they will suffer dire consequences. The goal of this book is to help companies develop key strategies for fraud prevention, and this prevention starts by first understanding the problem. In the next chapter, we will delve into criminological theory and look to create an understanding of the offender that is the focal point of our efforts.

CHAPTER 2

Criminological Theory

Criminological theory is the scientific study of criminal behavior; it focuses on the causes of crime, with the main goal of understanding the criminal mind as a road map for formulating effective preventative programs and policies. I have taught many criminology classes at the collegiate level, and their average duration is twelve weeks, so we will have to be very precise with the areas that we focus on in these next few pages! Criminology is an incredibly rich area and one that I find incredibly fascinating from a research and application perspective, so feel free to explore many areas on a deeper level on an independent basis. Criminological theory is applied in many criminal justice settings, such as homicide and robbery, yet there is virtually no application in the area of fraud. Understanding the motivation of criminals assists with developing highly focused strategies that will serve to deter criminality and reduce deviant activity. There are many internal and external factors that could have an impact on how a criminal behaves, and we will explore all of these in depth in this chapter.

First, we will examine many of the seminal criminological theories and learn about their foundational elements; we will start by exploring all three major schools of criminology: biological, psychological, and sociological. We will study how these theories have been applied in other criminal justice areas and specifically how they have been effective. Having an in-depth comprehension of the relative successes of these theories will help to assess their efficacy in a fraud setting. We will then revisit these theories and assess their specific application in insurance fraud, probing where they could be applied and have an impact in a counter fraud setting.

There is virtually no current application, or discussion, of theory application in insurance fraud, thus making this an area of uncharted exploration. At the academic level, there are very few peer-reviewed studies in insurance fraud and even fewer that make mention of theory application in this setting. Thus, this presents a very significant gap in the knowledge base, a gap that will be filled by discussions in this book. As we established in the first chapter, fraud is a crime that is growing at an alarming rate and thus in dire need of highly specific, focused strategies. Applying criminal theories to fraud will undoubtedly assist at developing these focused strategies. Criminological theory is an exciting area to investigate; we will discuss and explore biological, inborn factors, psychological, internal factors, and sociological, external factors. This exploration will also include relevant concepts such as motivation, deterrence, risk versus reward, and internal and external influence on human behavior. Let us start to dive into the mind of a criminal.

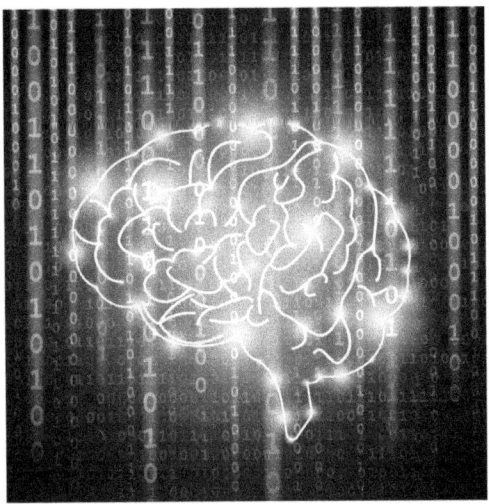

Schools of Criminology

Criminology is one of the most fascinating areas of criminal justice as it focuses on the behavioral aspects of crime. Have you ever wondered why some individuals have never received a speeding ticket, and yet others are not deterred after participating in repeat violent offenses? What is the difference between these individuals? Arguments have ensued stating criminals are born, while others focus on genetics and chemical imbalances, and yet others focus on environmental and sociological factors as the primary influencers of behavior. Criminologists study why criminals commit crimes and why they behave in certain scenarios. It is important to have this behavioral understanding, as then one can develop methods to control behavior, thus controlling crime. Criminology creates a global view of crime, allowing us to step back and see the bigger picture of deviance, and how important behavioral aspects are at predicting and preventing criminality. Criminological theory can be separated into three main categories: biological, psychological, and sociological. Each theory has been proven reliable and credible in the criminal justice arena. Let's delve into each theory and study its main tenants, highlighting the relative strengths and weaknesses of each approach.

Biological School of Criminology

Biological theory, also referred to as positivism or determinism, proposes that the primary root causes of deviant criminal behavior are abnormalities present within an individual. The main tenet of this approach proposes that behavior, including law-abiding activity, is determined by factors beyond an individual's control. The biological school of criminology focuses on genetic abnormalities, or defects, that are inborn and push an individual toward crime. Cesare Lombroso is a seminal researcher who pioneered this school of behavioral thought in the late nineteenth and early twentieth centuries (Edelstein 2016).

He portrayed criminals as people with certain inbred physical traits, such as poor body construction, sloping foreheads, asymmetrical skulls, long arms, and other apelike, subhuman characteristics, also referred to as atavism. Lombroso theorized that criminal behavior stemmed from these inborn factors, not from rational thought and free will. He found that almost one third of criminals fell under his atavistic classification.

Further research found compelling statistics that body types, or morphology, was also a strong predictor of criminal behavior. Somatyping is the study of the structure or build of a person to the extent that it exhibits signs of an ectomorph, a mesomorph, or an endomorph. William Sheldon was a seminal researcher in this area and found that mesomorphs (big bones and muscular shape) are more prone to committing violent and aggressive acts (Lilly, Cullen, and Ball 2015).

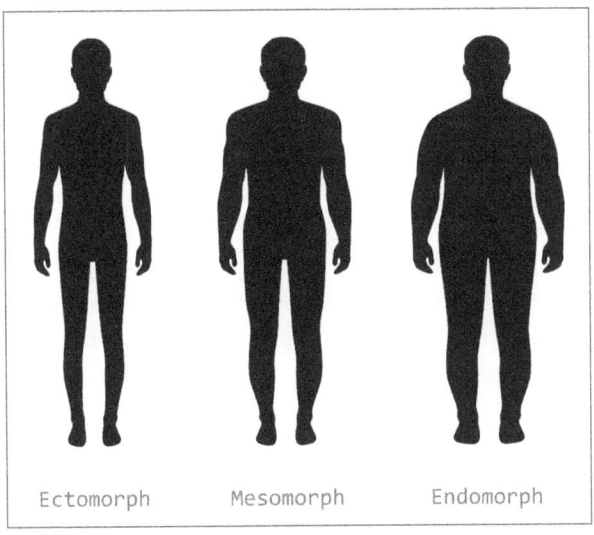

Ectomorph   Mesomorph   Endomorph

Other researchers within the biological school of thought postulated that chemical imbalances are a common trait in individuals that exhibit deviant behavior. Significant studies have been performed that establish a positive correlation between genetics and behavior. Seminal research shows a strong relationship between inborn chemical makeup and criminal behavior. There are many research articles in circulation that show compelling support that children with attention-deficit/

hyperactivity disorder (ADHD) are more likely to be involved in criminal behavior as they age. Other studies, such as one recently performed in Finland, revealed that out of 794 adult prisoners, an alarming 72 percent (568) tested positive for antisocial personality disorder (ASPD). The researcher further probed into biological factors and also affirmed that certain genetic, predisposed factors such as testosterone and cortisone have a direct impact on an individual's proclivity toward deviance (Chamberlain 2016).

I recall one conversation I had with a researcher during a recent training session that offers additional insights. As a trainer for Bureau of Alcohol, Firearms, and Explosives (ATF), I was conducting a session on counter fraud and had a very meaningful discussion with a medical doctor and researcher. We were discussing the genetic influencers of crime; he mentioned a study that he started several years prior but did not conclude. The subjects of the study were veterans who had been in Afghanistan and been in active combat. The study was looking to assess the relative testosterone levels in these ex-soldiers, and compare their levels to other nonmilitary personnel. His initial findings showed strong support that these combat-experienced soldiers had higher levels of testosterone and a higher degree of aggression. Parallel studies were also performed along gender lines, and these also revealed that the female hormone estradiol, equivalent to the male chemical testosterone, correlates to increased aggressive behavior in females.

We have established that there exists a very strong link between genetics and behavior. Multiple academic, reliable studies illustrate that there is a valid relationship between genetic inborn makeup and behavior; however, there appears to be a general lack of literature on what exactly triggers deviant behavior. If DNA analysis can pinpoint a specific genetic abnormality that is prone to criminality, how do we explain the law-abiding citizen that carries this gene? Further studies are needed to explore how sociological factors may cause the *stars to align* and trigger criminality. This will be a topic of discussion later in this chapter as we merge these applications into an insurance fraud context.

Genetic and congenital foundations of antisocial behavior

Contemporary research has suggested certain genetic and congenital foundations for antisocial behavior. Many seminal studies have been performed that show strong support that genetic abnormalities have a link to deviant behavior (Wiberg 2015). Specifically, a low variant of the monoamine oxidase A (MAOA) gene, when it is paired with negative childhood experiences or abuse, creates a high propensity for criminal deviance. This is such a widely accepted theory that the MAOA gene has been termed by the media as the warrior gene.

The MAOA gene controls the breakdown of serotonin and dopamine, which both play an integral part in mood and behavior; when there is a deficiency in the normal operation of this gene, imbalance occurs, which causes behavioral issues.

Genetic factors in victimization

Victim blaming, victim facilitation, and victim precipitation are very controversial issues within the criminal justice field. These alternative concepts imply that victims play a part in their own victimization, possibly by engaging in risky behavior. A significant number of research studies focus on the genetic makeup of the victim and propose that there is the possibility of the *born victim*, similar to the *born criminal*. These theories posit that *born victims* will entice their predators through some form of interaction. This application of genetics to victimization is one area of study that has been developing in recent years. This contemporary momentum was created by early foundational research, such as that performed by Allgulander and Nilsson (2000), that revealed strong support for these principles. In their seminal study, the researchers analyzed data from 1,739 victims of Swedish homicides looking for patterns of victimization. The resulting data showed certain patterns within the victims of these murders, but the most significant was the identification of previctimization brain injury as a risk factor in these cases. The researchers theorized that brain injury may cause irritability, agitation, belligerence, and anger, all deviant social reactions, which may contribute to victimization. The studies alarmingly revealed that there may be a correlation between one's proclivity toward victimization and genetic makeup. This genetic, or DNA, application to victimization is an emerging topic for exploration.

Certain research has illustrated that the dopamine and serotonin genes, DRD4, and specifically the 7repeat (7R) section of DR4R, has been linked to thrill-seeking behavior and attention-related problems, which traditionally have a negative impact on cognitive thinking. This reduction in cognitive ability results in an individual being more likely to engage in risky activities, in risky environments, and be less aware of environmental situations, which all facilitate victimization. All of the aforementioned are alternative theories to victimization; yet they show compelling links and strong correlations between genetics and behavior (Tiihonen et al. 2015; Vaske, Boisvert, and Wright 2012).

Brain dysfunction and criminal behavior

An additional research area that has shown powerful results lies in the discussion of how brain anatomy and dysfunction can be linked to serious criminal disposition and actual criminal behavior. One such seminal case study performed by Luukkainen et al. and Räsänen (2012) addressed the relationship between traumatic brain injury (TBI) and criminality in adolescents. The study revealed that adolescents that experienced any form of brain injury had a higher propensity to commit deviant behavior—53.8 percent with a TBI versus 14.7 percent without a TBI. The researchers found further supportive evidence when they delved into the categories of crime—42.9 percent TBI versus 9.1 percent non-TBI in violent offenses, and 29.4 percent TBI versus 6.8 percent non-TBI in nonviolent offenses. Overall, the study showed that a brain injury in one's younger years increased the risk of criminality 6.8-fold (95 percent 3.0–15.2).

Psychological School of Criminology

The psychological school of criminology focuses on the individual's mind—specifically, personality flaws and how they contribute to criminality. This line of thinking explains behavior as the consequence of individual factors, such as negative childhood experiences and inadequate socialization, which result in reduced cognitive development. Similar to the biological school, the psychological school focuses on the individual as the root cause of deviant behavior; environmental or other social issues are secondary, if considered at all. The merging of psychology and crime stemmed from early studies that attempted to determine the age of reason in delinquents—that is, the age at which an individual has the mental capacity to understand the difference between right and wrong. These initial studies forced researchers to analyze the mind of a criminal; an area that up until that point in time was underutilized in criminal theory application. Researchers came to determine that cognitive flaws and IQ deficiencies could be a strong predictor of deviant behavior; thus psychology was seen as a worthy area to insert into behavioral theory.

Psychological approaches are also referred to as psychoanalytic theories; no matter the title, the focus of this school of thought is on the mind of the individual. Classical Freudian beliefs are rooted in the psychological school of criminology—beliefs that claim disturbances in early childhood play a significant role in the tendency to engage in deviant behavior. Freud further explains these disturbances as being caused by the imbalance between the id and the ego. The id is the unconscious, irrational impulse that needs to be controlled in order for deviant behavior to be suppressed. This suppression is accomplished through the ego, which is the rational aspect of cognitive thinking, and the superego, which is the moral foundation of the mind. Freudian approaches claim that deviant behavior is the result of the conflict between the three competing psychoanalytic areas: id, ego, and superego (Lilly, Cullen, and Ball 2015).

Compelling research on personality traits shows strong evidence of their applicability within psychological theories of criminality. The most common technique to capture personality traits is by using the Minnesota Multiphase Personality Inventory (MMPI). The MMPI operates on a scale system and is widely used in many circles as a method to identify and assess personal, social, and behavioral problems. As this is a quantitative survey, researchers have been able to identify some key indicators of the psychopathic personality. Similar to a criminal profiling scenario, these scoring parameters are only an initial indicator of potential psychoanalytic traits; this is a starting point for researchers to investigate further.

Examples of criminals who fit into this psychological school of thought are Andrea Yates and Jeffrey Dahmer. Andrea Pia Yates was a Texas mother who was convicted and sentenced to life in prison for drowning her five children. The early works of Freud help to explain Andrea Yates's behavior in a psychological context. Freud argued that human nature includes a great number of instinctual drives, also known as the id, which demand immediate gratification. Prior to the murders, Yates had attempted suicide twice and had been hospitalized twice for mental illness, where she was ultimately diagnosed with schizophrenia (Cable News Network 2016). Serial killer Jeffrey Dahmer also showed early signs of mental disorder; at an early age he collected dead animals and showed signs of necrophilia. As postulated by many researchers in this area, individuals with a predisposition to crime or certain criminal personality traits are significantly more prone to engage in criminal deviance, which was the case with Dahmer.

Psychological theories of criminality can also be categorized into various mental disorders that contribute to deviant behavior. Conduct disorder is a psychological condition whereby an individual exhibits a persistent pattern of behavior that is not age appropriate. This is studied on a juvenile level and accordingly is thought to be indicative of later antisocial personality disorder. Juveniles diagnosed with control disorder exhibit less activity in the frontal regions of the brain during a risk-versus-reward scenario, indicating a psychological flaw in cognitive thinking and mental capacity.

Post-traumatic stress disorder (PTSD) is a mental condition whereby a traumatic situation causes later maladjustment and improper mental capacity. Military and law enforcement situations, child abuse, and being involved or witnessing a crime can all cause strong PTSD responses. Individuals diagnosed with PTSD have the inability to properly handle certain *normal* scenarios and as such may act in a deviant manner (Tibbetts 2015).

Sociological School of Criminology

Thus far we have explored criminal theory by looking at internal individual factors that affect behavior. The biological school focuses on inborn factors and how these genetic abnormalities cause deviant behavior. The psychological school discusses personality flaws and a lack of social development as the main driver of behavior. In contrast, the sociological school of criminology focuses on external, environmental factors as the key to behavioral influence. This school focuses less on individual microbehavior and more on macrogroup behavior. Crime is the final product of class struggle, inappropriate socialization, environmental factors, and the individual's location within the structure of society. The environmental school of criminology is an additional, but slightly lesser known, school of thought that can be applied in an insurance fraud context. Environmental criminology is very similar to the sociological school of thought in that they both place a strong emphasis on external factors and how they contribute to behavior. For purposes of this analysis, tenets of both the sociological and environmental criminological schools of thought will be included in this section of the analysis. These sociological and environmental factors will be the focal point of this book, as these external factors can be manipulated and therefore should be of great interest to the fraud fighter (Lilly, Cullen, and Ball 2015).

Routine activities theory

Routine activities theory is considered a highly credible theory in the criminal justice arena, and is ideally suited for application in an insurance-fraud setting. Cohen and Felson proposed routine activities theory in 1979 in an attempt to research environmental and external factors of crime. They were the first to argue that environmental factors significantly contribute to the nature and frequency of crime. Their significant studies contributed to this new approach on criminology; other theories, such as situational crime prevention and crime prevention through environmental design (CPTED), emerged as a result (Cohen and Felson 1979).

Early researchers of routine activities theory focused on the lifestyle changes of the 1960s, such as increased wealth and higher rates of participation in recreational activities outside of the home, which resulted in increased household theft and victimization by strangers. The foundation of routine activities theory proposes that three critical elements must be present in order for crime to occur: suitable targets, lack of capable guardians, and motivated offenders. When all three components are present, then crime will result (Tibbetts 2015).

Significant research on routine activities theory has been performed since Cohen and Felson introduced this new approach to crime causation. This research focused on applying the theory in various criminal settings and testing the three main components of target, guardian, and offender. These studies revealed extremely strong support for the main tenets of the theory. Early assessment of routine activities was performed by Spano and Nagy in 2005; the researchers focused on rural Alabama and found that increasing capable guardians reduces the risk of assault and robbery, thus providing strong support for the theory. Jackson, Gilliland, and Veneziano (2006) followed by studying college students in an academic setting and also confirmed that lack of guardians contribute to deviant behavior. Martin Andresen (2006) focused on crime in Vancouver; his study revealed positive support for routine activities as a viable explanation for auto theft, breaking and entering, and violent crime. Researcher Elizabeth Groff (2007) applied routine activities in a street robbery scenario and found compelling support for the theory. Contemporary assessment and application of routine activities has also shown positive results. Leukfeldt and Yar in 2016 found strong support in the cybercrime arena; Kula in 2015 showed solid findings for routine activities in analyzing the use of closed circuit TV for security applications in Turkey, and Warchol and Harrington in 2016 illustrated merit in the South African illegal wildlife trade!

Early and contemporary research has provided compelling support for routine activities as a viable approach to crime prevention in many different criminal circles (even illegal animal trade), yet, unfortunately, it is neglected in an insurance fraud setting. We will explore its application in a fraud environment in later sections.

Strain theory

Strain theory attempts to explain behavior by focusing on social strain or pressure. Individuals experience social strain when they feel pressure from being unable to obtain certain goals within the social structure. In American culture, we measure success by wealth, power, prestige, and material possessions; frustration ensues when a certain status (money, employment, school, or community) cannot be obtained; thus, frustration results, which is likely to generate criminal behavior. Strain theory states that crime stems from the conflict between people's goals and what means they use to achieve these goals; people feel strain when they cannot obtain these benchmarks of success through traditional means of employment and working. Thus, they resort to illegitimate (crime) means as an alternate to reach success (Lilly, Cullen, and Ball 2015).

Merton's theory on social structure and anomie serve as the foundational elements of classical strain theory. Merton argued that society maintains a balance between approved social means and approved goals, and proposes that there is sometimes a disconnect between the two as cultural ends can be obtained by illegitimate means. This research proposes that there is an incredibly strong American culture toward cultural ends, but not an equally strong culture of legitimate means. Merton viewed American values focusing more on monetary success than on the moral and ethical method to reach those goals. In this scenario, norms become weakened, which results in strain. Merton further proposed five logically possible adaptations that individuals undertake in reaction to this cultural strain. First, individuals conform, or accept their place in the social system and strive for success within their restricted means. Second, and the most common deviant reaction, is to innovate, or when goals are attempted through illegitimate means. Rebellion is the third response and occurs when an individual outright rejects cultural norms and means. The fourth is retreat and occurs when an individual withdraws and becomes a *dropout* by removing himself or herself from the goals and means dynamic. Lastly, ritualism occurs when an individual gives up struggle and focuses on what little gains have been made (Antonaccio, Smith, and Gostjev 2015).

Deterrence theory

Deterrence and rational choice theory are considered components of the classical approach to criminology. In the eighteenth century, Cesare Beccaria and Jeremy Bentham focused research on the legal aspects of crime and found surprising explanations toward criminal behavior. This classical approach to criminology follows the basic premise that the punishment should fit the crime, and, furthermore, actions are taken and decisions made by the free will and rational choice of the individual. This is a classical conundrum of risk versus reward and pleasure versus pain. This theory assumes that an offender will weigh the risks and rewards of an action and then take the most appropriate action based on a conscious decision (Carribine 2016).

Deterrence theory proposes that the punishment should be severe enough to outweigh the benefits, and if this is true, then an offender will make a conscious decision to refrain from deviant activity. Thus, if a legal system is unstable and does not provide adequate legal support and structure, offenders will be conscious of this and choose to manipulate the system and engage in illegal activities. A second element to deterrence theory proposes that punishment should be not only severe but also swift. If there is significant delay in investigating and prosecuting, than an offender will be more inclined to engage in deviant activity. The application of deterrence theory serves as both a general and specific deterrent. As a specific deterrent, the individual will refrain from further criminality if the punishment is swift and severe; as a general deterrent, the severe punishment of offenders serves as an example for others that may consider committing a similar crime (Lilly, Cullen, and Ball 2015).

Rational choice theory

Rational choice theory was developed in the 1970s and 1980s and carries many of the similar tenets of classical criminology. This theory furthers deterrence theory and proposes that offenders make a conscious, rational choice to commit or refrain from crime. Rational choice is closely correlated to the economic principle of a cost-benefit analysis; individuals weigh the advantages and disadvantages of an action and then make a conscious choice to act. Dissenters of this perspective feel that this theory assumes that individuals approach the criminal act with a profound calculation of the risks and rewards and deeply assess the pain-versus-pleasure dynamic. These skeptics illuminate research completed in the areas of property crime and theft that reveal that these criminals do not often operate using this rational thought process and are not deterred by severe sanctions. However, studies of white-collar criminals show that they are cognitively more advanced than other criminals and therefore have the capacity to operate on this deeper level of thinking; thus, there exists some merit for further exploration in an insurance fraud setting (Danchev 2016).

Jeremy Bentham and Cesear Becceria are two of the foundational minds that developed this line of criminological thinking. While at the Kilmainham Prison in Dublin, Ireland, the works of these two seminal researchers was evident as part of that prison's reform philosophy. The prison reform movement, which began in the late eighteenth century, was founded on the root tenets of rational choice theory. Many leaders within the criminal justice community of that time, and who were devout Catholics, felt that all men were sinners and convicts could only be reformed if they were brought to repent for their deeds while in prison. Bentham believed that all prisoners were motivated by their desire to seek pleasure and avoid pain, and thus all prison reform activities should focus on environmental factors that could be controlled in this pursuit. His line of thinking can be summarized as follows (This was prominently displayed at various parts of the prison.):

JEREMY BENTHAM

Men are rational, sensory beings who judge all things in terms of pleasure and pain. It follows that man's thoughts and behavior can be controlled through the senses. Government exists to create a system of rewards (pleasures) and punishments (pain) in a way that promotes the greatest happiness of the greatest number in society.

The function of prison is to create a totally controlled environment in which deviant behavior can be remodeled or reconditioned into socially acceptable behavior. And the method to do this is to design a new kind of prison (the Panopticon, or all-seeing eye) that controls all aspects of the prisoner's sensory experience. This includes separation cells, constant observation, and the rule of silence.

It is very evident in this early application of rational choice that controlling environmental factors is the root belief, and, furthermore, behavior modification will occur if these factors are manipulated in a positive manner.

Fraud Application of Theories

Thus far we have explored the three main schools of criminology and provided a summation of their foundational elements. We understand that all three schools have been applied in many criminal justice settings, and all have credibility. In the following section of the book we will discuss the practical application of these theories and what specific preventative approaches should be taken. We will see that of the three schools of criminology, the sociological school provides the most value for us operating in an insurance fraud setting, as we realize that environmental factors can be manipulated to alter behavior and create less opportunity for fraudsters. Our goal as fraud fighters is to reduce the vulnerability within our companies and agencies, and this ultimate outcome is achieved through reducing the opportunity to commit fraud.

Now that we have a basic understanding of the three school of criminology, you may be thinking, *How does this apply in an insurance fraud setting?* Or, *Which of these approaches can we apply and why?* This next section of the book will serve to answer these important questions by delving deeper into these theories and exploring specific applications.

Biological School of Criminology and Fraud Application

As discussed, the biological school of criminology proposes that behavior is driven by inborn factors—factors that are present at birth. Can the tenets of this theory be applied toward fraud preventative strategies? In order for us to answer this inquiry, we must analyze how this theory is currently applied.

The biological school of criminology focuses on genes and genetics as the main predictor of deviant behavior; preventative strategies would therefore focus on altering these genetics in alignment with law-abiding behavior. Genetic crime control strategies have been considered by researchers within the criminal justice field for many years. Studies have revealed that children diagnosed with attention deficit disorder can be effectively controlled by behavior-altering medication, prescriptions for stimulant and nonstimulant medications, antidepressant drugs, and high blood pressure medications.

The ethical concerns of a genetic crime control strategy are evident when we look at the history of its application. Traditional biological theorists, such as Darwin, addressed criminality as a form of biological flaw within the individual. The approach of blaming this *inferior* individual gained popularity, until the dangerous effects were illustrated in its application. The most significant occurred during the Holocaust, when millions were murdered as a result of being labeled as a member of an *inferior* class. There is no doubt that the serious implications of this application resulted in academics, researchers, and theorists readdressing its usefulness and application.

The contemporary approach of DNA mapping is another significant area of advancement within the criminal justice community, exonerating hundreds of previously convicted individuals (Olney and Bonn 2015). Researchers have begun to explore the ability to use DNA mapping to profile an individual and look to assess whether that person would be classified as law-abiding or deviant. This practice has received significant objection from the civil rights community and therefore is not seen as a viable preventative approach.

When we consider the above application of biological theories, we need to assess their usefulness in a fraud setting. It is unlikely that any insurance fraud professional will have the ability to genetically map an insured, claimant, or potential fraudster. Furthermore, there is no authority provided to insurance professionals to do so; therefore, the biological theories of criminology provide no practical application in an insurance fraud setting.

Psychological School of Criminology and Fraud Application

As presented earlier, the psychological school of criminology focuses on personality flaws within an individual and the mental capacity, or incapacity, of an offender. Psychological theorists propose that deviant behavior is driven by negative childhood experiences and by how these may create cognitive flaws that result in improper social development. These theories, and a significant element of this school of thought, are grounded in the works of Sigmund Freud. Freud's seminal works on psychodynamic principles focus on how human behavior, specifically deviant behavior, is largely driven by unconscious aspects within an individual's mind. Freud focused on how early childhood experiences are one of the main drivers of adolescent, and even adult, behavior. His studies centered on the conflict within the human personality segments of the id, ego, and superego (Myers and Dewall 2015). Furthermore, improper psycho development at various younger stages significantly impacts an individual's ability to operate in a nondeviant manner as an adult. Freud felt that aggression was present in all individuals; however, in adults that had proper experiences in childhood, these aggressive tendencies could be suppressed. The complication arises when maladjusted individuals are faced with aggressive impulses; Freud argues that these individuals will not be able to suppress this aggression, and deviant behavior is the result.

Before we consider if these psychological theories can be applied in a fraud setting, we must address how they have been applied in other areas. As discussed, these theories focus on the maladjustment that occurs in one's younger years. Policies and practices that integrate psychological theories focus on the need for both short- and long-term family and individual therapy sessions, which could include individual therapy, group therapy, and family counseling. All these approaches would focus on guiding interaction with the goal of getting to the root cause of maladjusted behavior. We know that psychological theory focuses on experiences in younger years, so these therapy sessions would attempt to isolate specific instances of impact and then psychoanalyze these situations and look to reassess the action taken. This is a highly cognitive therapeutic approach and one whereby a highly trained therapist is required.

As we consider the applicability of psychological theory in fraud policy, we must ask if these therapeutic approaches are possible in a counter fraud scenario. No matter whether we are employed by a public agency or private company, it is not plausible that these entities would have any interest, or resources, to set up and administer a therapeutic environment for insureds, claimants, or other participants. Furthermore, a company or agency would also have to consider if these programs would be implemented on a universal or selected basis, all of which requires time and resources, which again, does not seem feasible. In conclusion, it appears that psychological theories have no applicability in a fraud setting and thus should not be considered as a feasible portion of a counter fraud program.

Sociological School of Criminology and Fraud Application

Thus far we have established that there is limited application in fraud prevention using the biological and psychological schools of criminology. We will now further delve into the sociological school of criminology and assess its application in a fraud environment.

The next section will discuss the four sociological theories discussed previously and apply them specifically in an insurance fraud setting. It is important to note that there are dozens of sociological theories; however, the four chosen are the most applicable in an insurance fraud scenario for reasons to be discussed.

Routine activities theory and insurance application

As presented earlier, the research confirms that routine activities theory is a valid and legitimate approach toward crime prevention strategies in many crime areas, yet how can we use its foundational elements to help fight insurance fraud? We understand that when all three components of routine activities are in alignment (suitable target, lack of capable guardian, and motivated offender), then crime is the result. Therefore, strategies to prevent crime must focus on developing one of those three key areas: suitable targets, capable guardians, or motivated offenders. It follows that if one of these three areas is manipulated, then the connection between them will be disrupted and behavior (crime) will be avoided.

The research in other disciplines, as discussed earlier, has revealed that increasing the presence of capable guardians into a scenario disrupts the occurrence of crime; therefore preventative fraud efforts could focus on increasing the visibility of these individuals. A capable guardian in an insurance fraud setting could be anyone that is a key gatekeeper or point of contact in the process, someone that has the unique opportunity to act as an authority figure within the claim and investigative process: loss report takers, insurance agents, claims adjusters, field appraisers, claim managers, analysts, police officers, prosecutors, special investigative unit investigators, state fraud investigators, and many others.

One early and very seminal study on claims fraud (Tennyson 2008), argued that claims/fraud/law enforcement investigators are in very critical roles, and their mere presence has a significant impact on reducing fraudulent claims. If we apply what we have learned in regard to routine activities, these investigators would undoubtedly be considered gatekeepers, or capable guardians, in this context. This study alludes to how insurance companies often overlook the deterrent effect that these capable guardians have on the filing and follow-through of a potentially fraudulent claim. The research further showed that the most effective prevention procedures involve those derived from deterrent strategies from these key individuals. While at a recent international fraud conference, I asked ten top fraud-fighting executives to participate in a fraud consortium to develop current trends and strategies in the market. One of the most interesting insights that resulted was how the increasing prevalence of automated claims processing systems without strong human oversight could be one of the main driving factors for increases in fraudulent activity. The consortium felt that the use of technology was absolutely vital for the future success of insurance carriers; however, they warned that there still needs to be some aspect of human supervision to these automated systems. There is no doubt that fraudsters can *feel* the lack of human intervention, which will push them into action as they capitalize on this vulnerability. Companies, therefore, need to complement the automated systems with human intervention in order for fraud detection to be truly effective.

As fraud fighters we also face the disturbing statistics surrounding the general acceptance of this crime, with the research illustrating that on average 25 percent of the general public believes that it is acceptable to pad an insurance claim for additional gain. Later in this book, we will conduct an in-depth analysis on the psychology of entitlement and how this phenomenon has caused significant increases in the occurrences of fraud. Policyholders will justify committing fraud as they feel that they are *owed* all of the premium dollars that they have paid their respective insurance companies through the years. Further research has also shown that one of the most significant aspects of criminal prevention is the organizational structure of key guardians, capable guardians, supercontrollers, or gatekeepers, which are in strategic positions to assist with crime prevention. As was discussed earlier, these key guardians could be anyone involved in the claim or investigative process, starting with claim reporting/sales employees to anyone else involved up to the final disposition of the settlement. The analysis illuminates how crime prevention strategies should focus on mobilizing these gatekeepers and teaching them to be more effective in their roles in the routine activities perspective; these key individuals are seen as the nucleus of crime prevention strategies.

Thus far we have seen how capable guardians are a vital part of an effective fraud strategy, which of course is a foundational element of routine activities theory. Suitable targets are another element of routine activities that bears brief discussion. It is vital that agencies and companies have an understanding of the current counter fraud environment and what specific strategies one's peers are utilizing. Having this critical competitive information is vital for the success of a preventative fraud strategy. The criminal element will focus on companies and agencies that are weaker than others, as these are easier targets and offer the most reward with minimal risk. Therefore, it behooves a fraud department to develop strategies that help them avoid being seen as a soft target among one's peers. Agencies should focus on making themselves unsuitable targets, which studies have shown will divert criminals to other suitable targets.

As discussed, the majority of the research on routine activities theory proposes that there must be a presence of capable guardians for crime prevention strategies to be truly effective. Researchers in this area argue that criminals must feel an awareness of capable guardians, or crime will ensue; guardians must make themselves visible in order to thwart the appearance of vulnerability.

The application of routine activities in insurance fraud is uncharted territory in need of further study to assess its applicability and effectiveness. Studies in parallel areas of criminality have revealed that this theory is a credible and legitimate approach toward reducing opportunity, which is a key component of deterring deviant behavior. Routine activities has great promise because insurers can manipulate some aspects of the three components. As we are aware, there is limited application in fraud scenarios; thus, we need to draw upon parallel areas to assess the validity of these theories. One such early parallel study tested routine activities theory in a street-robbery setting (Groff 2007) and found extremely strong support for the tenets of this concept. The crime of street robbery was chosen because it involves a direct form of interaction between a victim and an offender, and the crime is one for economic gain; both factors likely will result in a rational choice decision. This particular study is valuable to us in the fraud industry as it focuses on the relationship between the victim (suitable target such as insurer or consumer) and offender (motivated offender). A second commendable study performed by Leukfeldt in 2016 tested the merits of routine activities by focusing on cybercrime. In this multivariate analysis based on a large sample size of $N = 9{,}161$, the author showed that visibility clearly played a role in cybercrime victimization, adding credibility to the argument that organizations must increase counter fraud presence (Leukfeldt and Yar 2016). Making alterations to help reduce the chances for victimization as outlined in routine activities was the key to reducing criminal opportunities, according to these studies. In applying this to insurance fraud, if insurers as suitable targets alter their anti-fraud strategies, then this will reduce opportunity, disrupt the continuum, and reduce fraud occurrences. Some specific ways that insurance companies can disrupt the continuum and make themselves less-suitable targets would be to make more-aggressive red flag systems, increase the use of critical gatekeepers such as analysts and claims staff, increase training and awareness to develop stronger gatekeepers, use technology as a gatekeeper, and support public outreach.

The research cited thus far has shown support for routine activities as an effective means to explain and deter criminal offending. Any manipulation of victim (insurer) activity reduces the opportunity to victimize. This theory has significant implications for anti-fraud strategy, as the research supports a fraud-reduction system focusing on making the target harder for an offender to victimize. As discussed, sociological theories have significant merit in insurance fraud applications, as they focus on environmental factors that can be manipulated. It is worthy to briefly discuss contemporary perspectives of environmental criminology in order to understand the value of environmental manipulation and behavioral modification.

One such contemporary environmental theory is crime prevention through environmental design (CPTED). CPTED theorists focus on the environmental factors of crime by analyzing extraneous factors that can be altered to help reduce crime and victimization. The foundation of CPTED began when proponents found that altering the physical properties of buildings and other physical elements caused a significant reduction in criminality (Jeffery 1971). In 1971, criminologist C. Ray Jeffery published the book entitled *Crime Prevention through Environmental Design*, but the principles did not gain ground until the late 1990s and early 2000s. Early application in an urban-housing setting in Sarasota, Florida, in 1999 and 2003 showed that CPTED strategies are highly effective at reducing criminality (Carter, Carter, and Dannenberg 2003). Criminal justice professionals in Sarasota reduced opportunity by modifying environmental factors such as zoning regulations, buildings, and other physical structures. This basic theoretical foundation grew into a mind-set that modifying environmental factors provides the rational offender less opportunity to commit criminal behavior, thus reducing suitable targets, one of the main components of routine activities theory (Cozens and Love 2015).

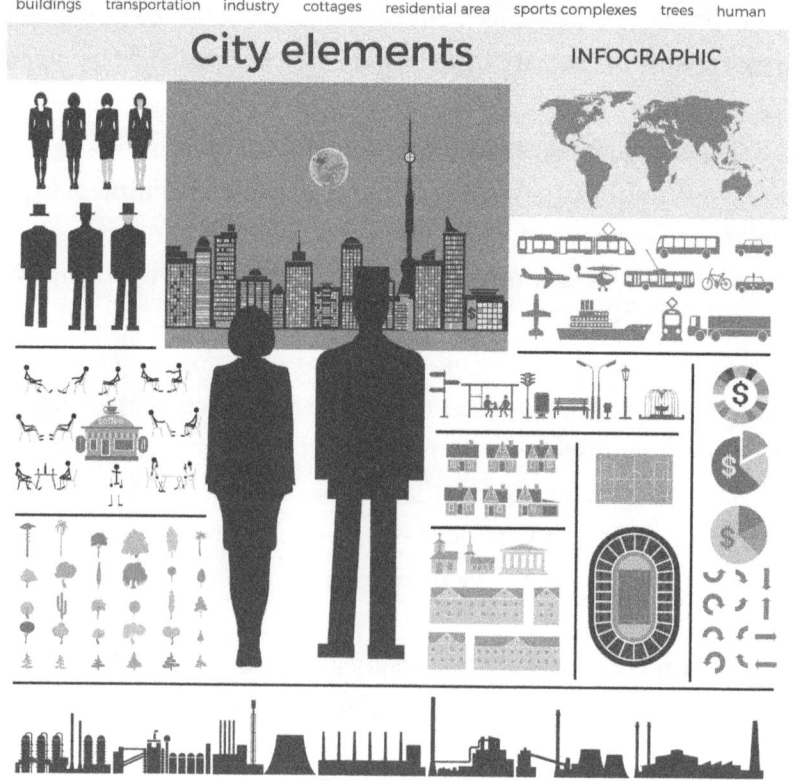

An important CPTED strategy involves the concept of foreseeable danger; in this critical step, a risk assessor analyzes the case, crime, and company, and then determines potentially vulnerable areas on which to focus efforts. This risk assessment is comprehensive and could entail reviewing prior data, spreadsheets, internal company information, and external trends. Social environmental theory proposes that crime is not a matter of motivation, but of opportunity. These rational-choice theories focus on how situational or environmental factors contribute to crime causation, and are based on rational thinking. Potential offenders use choice-structuring properties—the crime's cost-benefit analysis. Making fraud more difficult will result in less opportunity and fewer fraud occurrences.

The reduction of opportunity has been presented as one of the main lessons of this book. As such, it is also important to consider the perspective of the victim when considering the precipitation of crime. Victimization research illustrates that many offenders do not premeditate crime; they act when an opportunity presents itself. Past interviews with convicted criminals have revealed to us that many stated that their criminality was never premeditated; it just happened. A significant amount of research reveals that offenders act in a deviant manner only when they are presented with an opportunity; thus it benefits insurers to develop as many ways to reduce this opportunity as possible to thwart fraud events.

Counter fraud strategies to reduce opportunity could include the following:

- more aggressive up-front identification of potentially fraudulent cases, such as at the point of sale
- including sales and agency staff in counter fraud efforts
- effective software packages to identify high-risk claims
- development of gatekeepers, such as analysts or SIU/claims liaisons
- coordination with underwriting staff to assist with fraud identification

- aggressive investigation once a potential fraudulent case is identified
- showcasing cases where fraudsters have been prosecuted
- tighter underwriting standards that align with fraud trends
- use of network visualization to identify patterns
- use of predictive modeling to help identify vulnerable areas
- development of red flags in order to communicate trends to others in the company/agency
- coordination with other departments such as finance and marketing regarding fraud identification
- training of staff on fraud identification and increasing awareness of the fraud problem
- partnering with law enforcement and other investigative groups in regard to emerging fraud patterns

Strain theory and insurance application

Strain theory is relevant to the anti-fraud community because many opportunistic fraudsters commit fraud to help alleviate social strain. In any tight job market, where reductions and downsizing are quite common, many people do not have the traditional means to earn a stable income and obtain their benchmarks of success; they commit insurance fraud to fill that void. Of particular interest is the application of strain theory to the organized and opportunistic fraudster.

Robert Merton and Albert Cohen furthered general strain theory (GST), and focused primarily on lower-class individuals and how they resorted to crime in order to resolve the strain they felt in achieving goals through normal means (Cohen and Felson 1979). Classic strain theory focused on how the lower class lacked the money and prestige of their middle-class peers and as such had to resort to criminal behavior to achieve this desired middle-class status. This perspective could be applied to the opportunistic fraudster who may be from a lower-class demographic, but what about the fraudster who is in the middle- to upper-class tier, who seemingly has basic financial needs satisfied yet still resorts to fraud? Although not a central point of Merton's work on GST, he also argued that middle- and even upper-class individuals can also feel the same strain as lower-class individuals. He presented the argument that these individuals experience even higher levels of strain than lower-class individuals, as there is increased pressure and stress to achieve higher levels of success in the upper classes. In addition, once a certain lifestyle is achieved, which could include exotic vacations, houses, and vehicles, there is incredible peer pressure to maintain this level of economic success. Becoming a member of this exclusive, elite *club* can be quite addicting and drive middle- and upper-class individuals toward fraud in order to continue to maintain this level of living.

An interesting psychological phenomenon is also occurring among our younger generation, a phenomenon that drives fraudulent occurrences and confirms the credibility of strain theory. In my fifteen years of teaching at the collegiate level, I have seen a very disturbing trend in our younger generation and their feeling of entitlement. I have seen a slow increase in the degree of entitlement that this younger demographic expects from society. This became particularly evident one summer evening when three of my fellow PhDs and I were walking to our vehicles after teaching an evening class. We recognized several of our students entering their vehicles in the parking lot and immediately noticed that all three had new, luxury, exotic vehicles, automobiles that seemingly were out of their demographic capabilities as first-year students. As PhDs, we, of course, had to further delve into researching this seemingly simple observation. We investigated and came to understand that none of these students *earned* the vehicles in the sense that they had the job, income level, and/or credentials to afford them. All three students leased their vehicles; two students did not work and lived at home with their parents, and the third had a small part-time job and lived in an apartment, but devoted almost 75 percent of his earnings to pay for the lease of his luxury vehicle. When we conducted informal interviews of these three students, all had a very distinctive sense that they earned the right to drive these cars because *they worked real hard*. Yet, in our opinion, their careers were just beginning!

This is not to imply that all those in the younger generations, and all students, have this same mentality, as I have worked with many students who have an incredible work ethic. Yet it is worth noting that I have noticed an increasing trend in the sense of entitlement in the students that I have taught throughout the past fifteen years. This has very interesting implications to us operating in the fraud industry, as those in these younger generations may resort to fraud in order to fulfill class gaps and satisfy their sense of entitlement as they cannot achieve lifestyle goals through normal means; this could be a trend worth monitoring.

Deterrence theory and insurance application

Deterrence theory is based on the classical criminological works of Cesare Beccaria and Jeremy Bentham, who portrayed criminals as individuals that can be deterred from deviant behavior by the modification of external or environmental factors. This school of thought also assumes fraudsters are rational thinkers who will avoid criminal behavior if they are highly deterred (Carribine 2016). It is important for companies and agencies to focus on actions to increase deterrence on both a specific and general basis. On a specific level, fraudsters must be taught to think that fraud will not be tolerated, which will hopefully result in behavior modification in a positive manner. Specific deterrence focuses on ways that a particular individual can be deterred from a certain activity—in this case, fraudulent activity. On a more global level, companies and agencies must showcase that they are hard targets and thus deter fraud on a higher operating level. General deterrence focuses on ways that companies can send a general message that their company or agency will not tolerate fraud; this will hopefully deter criminals before they even consider action. One strategy is to develop and publicize strict zero-tolerance programs by making it clear that fraud cases will be diligently investigated and prosecuted. Insurers can also support and publicize fraud convictions, which will send a strong message to future potential fraudsters. If we assume that insurance schemers are rational thinkers, then they will respond to deterrence if convictions are highly publicized.

Legislative anti-fraud efforts also will help create a deterrent effect. Because fraud is traditionally underprosecuted, this sends a message to the public that it is a relatively low-risk crime. There is no deterrent effect when the punishment is low because the reward will be worth the risk to the rational fraudster.

Rational choice theory and insurance application

Rational choice theory is one of the foundational perspectives of sociological and environmental criminological principles. It is based on the premise that offenders make a rational choice to commit crime, and are influenced by environmental factors. People have certain preferences and are motivated by wants and needs, which determine the choices that they make. These rational individuals will weigh the costs and benefits of an action and then make a conscious choice as to the result that will provide them with the most satisfaction and reward. The key word when considering the application of rational choice is that the subjects must be rational. It is important to consider this in application, as irrational individuals often do not consider the risks and rewards of an action and therefore will not be affected by the foundational principles of rational choice theory.

As with the biological and psychological schools of criminology, there is no significant application of sociological theories in an insurance fraud setting. Therefore, in order to assess its validity as a theory, we must look at how it has been applied in other research areas.

Intriguing research applying rational choice theory in the area of terrorism has revealed interesting results that show strong support for the theory in this criminal field. Public opinion surveys have shown that the public characterizes terrorists as psychopathic and extremely mentally unstable individuals. However, professional interviews with terrorists have revealed quite the contrary; they do not normally exhibit signs of psychotic disorders, such as schizophrenia, anxiety, or insanity (Harrison 2015; Schneider, Brück, and Meierrieks 2015). Thus, a rational choice application to terrorists could be extremely useful at attempting to explain their behavior, as we know that most are capable of making rational choices. We know that terrorists have a specific political agenda; applying rational choice theory would explain why they have realized that various forms of violent terrorist acts are the best option at fulfilling these goals.

Most criminals operating in the white-collar crime arena are more cognitive than other criminals, thus making them more likely to participate in a cost-benefit analysis as they consider participating in a deviant act. Rational choice theory would then be ideally suited for application in a fraud setting, as we know that these individuals are more intelligent and *consider* fraud before the actual act is committed. It logically follows that fraud preventative efforts should focus on ways that draw fraudsters away from participating in fraud in the first place. It is the central premise of this book that counter fraud efforts should focus on ways to reduce the opportunity to commit fraud; our focus should be to make a potential fraudster think about the act and then decide not to participate due to outweighing using the risk/reward perspective.

There are two main perspectives that fraud fighters should consider as elements of rational choice are applied; one focuses on the global crime of fraud, and the second centers around fraud within their respective companies and agencies. We are aware that fraud has become the preferred method of crime for organized criminal groups and terrorist cells that are drawn to the high reward and low risk of fraud as compared to other forms of crime such as drugs and extortion. These groups prefer fraud, as there are huge financial gains and relatively no risks. Traditionally, these criminal organizations have participated in drug activity to fund their habits, which can be extremely violent and unpredictable. Fraud is a very enticing criminal option, as it does not involve a violent element and also has very lucrative payouts. And these organized fraudsters are cognitive and maintain a very contrived approach to committing crime perfectly suited for a risk/reward decision-making process. Thus, in this global perspective, counter fraud efforts need to focus on ways that communicate to these organized elements that fraud is not worth the risk as compared to other forms of crime. This can be accomplished by showcasing cases where agencies and companies were very aggressive during the investigation, prosecution, and punishment phase. In colloquial terms, we need to send a message to the criminal element that fraud will not be tolerated; the punishment is not worth the benefits. If we recall, fraudsters operate on a higher cognitive level than other criminals and therefore can make conscious decisions to refrain from an act if the costs outweigh the benefits. Our counter fraud efforts should therefore focus on ways to communicate our investigative activity and successes. This can also be accomplished by investing in public awareness campaigns and media outlets that focus on our activities. All these techniques will serve to apply rational choice at the global level.

Counter fraud efforts also need to consider a more macro perspective—that is, focusing on ways that their respective companies and agencies can avoid being seen as a soft target among their peers. Fraudsters admit that they target those companies that have weak fraud prevention systems, significantly capitalizing on these vulnerabilities. I recall one field investigation I was involved in early in my career where the subject of the investigation was a questionable medical facility that was being run by an organized crime ring. At the conclusion of the investigation, and after arrests and search warrants issued, there was a very interesting find in the rear of the medical facility. On a large white board in the back meeting room of the business was a chart that listed insurance company names, claim representative names, and then the words *hit* or *no hit*. The facility kept records of which company, and claim rep, was the easy or soft target!

Fraudsters are comparable to other criminals in the sense that they prefer to target these weaker, more vulnerable victims that are seen as easier prey. This is not to imply that our counter fraud community should act in a manner that is only self-serving to each of our respective companies, but to only bring to light this mentality of the criminal element. Thus, it is important to attend conferences and meetings and network with others in our industry to assess our level of vulnerability among our peers. This will serve as a benchmark as we develop new programs and policies to assist with fraud detection and prevention. If we are not seen as the most vulnerable target among other parallel companies, this will undoubtedly deter many fraudsters from targeting us when, and if, the opportunity to commit fraud arises.

It is apparent that rational choice theory has definite application in a fraud setting. As outlined above, there are many aspects of the theory that can be applied to assist with reducing the opportunity to commit fraud and ultimately affect the risk/reward balance in our favor. We understand that fraudsters operate on a higher cognitive level than other criminals and therefore can definitely be affected by our efforts to increase the risk and reduce reward.

Criminology is an extremely exciting area to explore; this field helps us understand human behavior, and specifically criminal behavior. In this chapter, we explored the three main schools of criminology: the biological, psychological, and sociological approaches, and learned about their foundational elements. We further analyzed each school and realized their application in a fraud setting, specifically coming to the understanding that the sociological school has the most merits in our industry. Rational choice, routine activities, strain, and deterrence theories help us to understand a more global view of fraud prevention, a view that considers the psychology of the fraudster. In the chapters that follow, we will take all of these theories and apply them in specific anti-fraud scenarios.

CHAPTER 3

The Fraud Triangle

The fraud triangle was first introduced in an article written by early criminological researchers Donald Cressey and Edwin Sutherland and later formalized by Steve Albrecht. It is universally applied in many settings and is considered one of the seminal methods to analyze fraudulent issues. Sutherland is a well-known researcher who developed differential association theory, which proposed that criminal behavior is connected to an individual's association with the criminal environment. The main tenet of this theory focuses on how each of us is faced with many social encounters during our lifetimes, and some of those situations will occur with individuals who have criminal tendencies (McMahon et al. Bressler, and Bressler 2016). It follows that this association to the criminal element in this social setting will push an individual to eventually engage in criminal behavior. Sutherland and Cressey believed that

- intimate personal connections are the main drivers of learning criminal deviance;
- verbal communication, and other forms of interaction, is also a significant driver of learned criminal behavior;
- deviant behavior is learned from social interaction and not inborn;
- through association, one learns not only about a specific crime but also about the relevant motives, tactics, and techniques; and
- criminals will rationalize the crime, which will strongly outweigh the risks that accompany criminal deviance.

Sutherland and Cressey did not develop the specific fraud triangle that we apply today, but they did initially propose the concepts of rationalization and opportunity, both core elements of the fraud triangle. The fraud triangle consists of three main areas: pressure, rationalization, and opportunity.

The fraud triangle is based on the central hypothesis of Donald Cressey:

> Trusted persons become trust violators when they conceive of themselves as having a financial problem which is non-sharable, are aware this problem can be secretly resolved by violation of the position of financial trust, and are able to apply to their own conduct in that situation verbalizations which enable them to adjust their conceptions of themselves as trusted persons with their conceptions of themselves as users of the entrusted funds or property. (McMahon et al., Roden et al.)

The fraud triangle

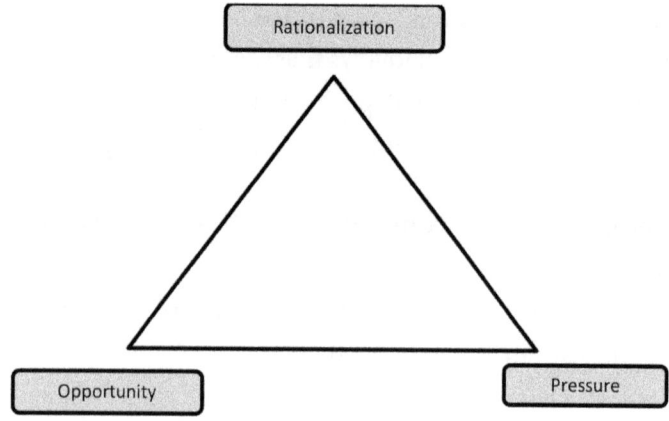

Pressure

Pressure can come in the form of financial difficulties, debt, or greed, all of which create an incentive, or push, to commit fraud. This area of the triangle attempts to explain what caused the crime to precipitate in the first place; an individual experiences some sort of pressure and then realizes that this problem cannot be resolved using normal, legitimate methods; thus deviance becomes a viable option. It is important to note that this pressure can come from a personal perspective, such as an individual that is in personal debt, or a professional perspective, such as an individual whose job or career path is challenged. Interviews performed on convicted fraud offenders have revealed that they often report increased job pressure and stress, both strong possible contributors to increasing their tendency to commit fraud.

Declining economic factors in areas of employment, job growth, expenses, and other areas have put additional pressure on many individuals; thus, as a culture, we seem to be exposed to this pressure on a daily basis. This has serious implications for those of us operating in the anti-fraud industry, as this constant pressure undoubtedly increases the incidents of fraud. Interestingly, one of the main findings of my dissertation study was that the bad economy contributes to fraud.

My qualitative dissertation research study was entitled "A Phenomenological Study of the Challenges and Barriers Facing Insurance Fraud Investigators," and consisted of fifteen in-depth interviews from a credentialed sample of insurance fraud professionals. Over five hundred pages of text resulted from the interviews, which was then transcribed and analyzed using Moustakas's data analysis approach and Atlas.ti.7.0, a data analysis software package. One of the main research questions posed to the sample asked for their opinion on the main trends in insurance fraud in today's environment. The thematic analysis of their responses revealed the following four themes:

1. Increase in soft/opportunistic fraud.
2. Increase in medical/no fault fraud.

3. Fraud is becoming more dangerous and sophisticated.
4. The bad economy contributes to fraud.

Some direct quotes from the expert sample:

- "There is an absolute, direct relationship between what is happening economically and our quarterly and annual fraud referral numbers."
- "As the economy drops, the opportunistic fraudster seems to take more opportunity."
- "There is so much joblessness, people are upside down on their mortgages; everyone still wants a piece of the pie. You used to get ahead by working for a company for your entire career; you always had a job, not anymore. A company will quickly drop you, and then you are looking for options, and there is one option that is very easy for everybody, and that is insurance fraud."

These financial difficulties also impact businesses that must also look for new methods to maintain profitability in the difficult economic market. Often, these methods and tactics can lead to increases in fraudulent activity. One such trend appears to be in auto loan financing whereby financial companies are significantly lowering their approval standards for auto loans and leases. There have been many reports of unqualified individuals being approved for loans that they would not normally be qualified for, as they are termed extremely high risk, yet are being allowed to purchase and lease vehicles. This has caused increases in the frequency of owner give-ups, which occur when the owner realizes he/she cannot afford to make payments on a vehicle and helps to stage an auto theft or car fire with the goal of collecting the insurance money.

Personal pressure to commit fraud can come from economic hardships being faced by our population; these hardships cause challenges in making payments for our cars, homes, college tuition, business expenses, and daily expenses. These negative environmental factors contribute to increases in fraudulent activity in the insureds, claimants, and policyholders that we come into contact with. Yet we must also consider the exposure we face internally, from our own employees.

Employee theft can cause a significant impact on the financial status of an organization. According to the Association of Certified Fraud Examiners, US businesses lose approximately 5 percent of their annual revenue to employee fraud; even more alarming, fewer than 10 percent of these incidents are discovered! Recovery statistics are equally as disturbing; most employers recover only 20 percent of the original loss, and 40 percent of the time, they recover nothing at all (Association of Certified Fraud Examiners, 2016). A recent report that examined recent federal court cases of employee theft provides some very insightful statistics in regard to this unique crime:

- The most common type was straight theft of funds (38 percent).
- The second most common type is check fraud (34 percent), which occurs when an employee alters, forges, or makes a check payable to himself/herself.
- Women were offenders in 60 percent of the cases, so they were considered a higher risk.
- The median loss for schemes carried out by women was $243,447, 30 percent less than males.
- Males were more likely to commit vendor fraud, which was defined in the report as "a perpetrator diverting employer funds through the creation and submission of false invoices issued by fictitious companies."
- Median age was fifty years old.
- Perpetrators most often worked in a senior level position and had a long, positive relationship with the company.

During my twenty-two years working in the anti-fraud industry, I have worked with some of the most talented, loyal, and dedicated professionals that exist; I doubt there are individuals with such stellar characteristics working in any other industry. Yet during those twenty-two years, I consulted with several companies where a serious occurrence of internal theft occurred; both of the incidents were entirely separate, yet both had extremely similar circumstances and characteristics. The two offenders were males in their midthirties who held bachelor's degrees, had approximately fifteen years of professional experience, and held positions at the upper operational level in the investigative department. Neither of them had any prior criminal history, and they were by all standards *ideal* employees who consistently scored above average on their performance evaluations. Yet both ended up being investigated, arrested, and prosecuted by law enforcement for theft-related charges.

One of the offenders was creating fictitious claims and issuing payments to fabricated body shops and claimants, all the while collecting the checks he issued from these fake companies and individuals. The second offender had a very similar modus operandi and was issuing checks on legitimate claims to fabricated claimants, claimants who he had a relationship with and could collect a kickback from. One of the offenders collected approximately $55,000 and the other $70,000. After both offenders were arrested and prosecuted, further details pertaining to their personal circumstances were released, which allowed us insights into the potential motivating factors behind their activity. Both of them had experienced a form of financial pressure in the year preceding the theft; one had a spouse that lost a job, and the other lost a significant amount of money from a family business. In several postprosecution interviews, both mentioned that they felt an incredible amount of pressure to maintain their current middle-class lifestyle, and both needed to take action before further financial loss ensued.

Internal employee theft should be a priority for companies and agencies. In later chapters we will address specific strategies to tackle internal theft—strategies that focus on internal controls and methods to help avoid and identify these incidents. Pressure is one of the main drivers of employee theft as illustrated, and is one of the main components of the fraud triangle. Now, let's turn our attention to the second main component: rationalization.

Rationalization

Rationalization is the second component of the fraud triangle and focuses on how the fraudster justifies his/her behavior by minimizing or making himself/herself *feel better* about committing the act. Many opportunistic fraudsters have no criminal history and are first-time offenders; they view themselves as honest people who unfortunately are faced with negative circumstances. Almost all fraud incidents involve some sort of rationalization, which could include statements like the following:

- The company owes me.
- I am just borrowing the money.
- It is not really hurting anyone.
- Just this one time.
- They owe me for all of the premiums I have paid.
- Insurance companies are rich; they won't miss a little money.
- They (insurance companies) are trying to get one over on me with my claim, so I am going to get them back.
- I am entitled to this.

We will explore some of these responses in more detail in the next section, but all quotations focus on how the fraudulent act can be justified, rationalized, and even borderline acceptable by the fraudster in certain situations. The rationalization can be centered around the specific company—"They owe me for all of the premiums I have paid"—or more globally focused on external factors. In other words, an insured or claimant could be faced with a job layoff, or some external situation that didn't involve the insurance company directly, but his or her insurance company becomes the victim of the fraud. For example, if an insured or claimant was fired from a job and became disgruntled as a result, he or she would justify committing a fraudulent act based on his or her mistreatment from the employer. Rationalizing an act makes people feel better about committing fraud, and in their perspective, they are still honest people.

As stated earlier, the research reveals that approximately 24 percent of the public believes that it is acceptable to commit fraud to make up for a deductible, and 18 percent felt it was okay to increase the claim amount to make up for premiums paid. An overwhelming majority of the sample participants in my dissertation study felt that fraud was a larger social issue, an issue whereby the general public has come to accept fraud as part of doing insurance business; this became one of the main themes of the study. One of the dissertation participants remarked, "We have a public barrier, the feeling that it is okay to commit fraud because I have paid my premiums, and I am owed this; this definitely has to change, you have to show people that there are damaging effects of fraud, both financial and physical, in order for change to come."

Multiple sample participants discussed how many fraudsters justify their actions because they feel that they are owed and entitled to a certain lifestyle: "These claimants are unemployed or underemployed and still want all the toys (cars, boats, etc.), and they turn to fraud as part of their income."

Instances of rationalization can also be the result of insurance companies themselves precipitating this mentality. The majority of the dissertation sample felt that insurance companies are their own worst enemy because many of them create the public opinion that they are extremely profitable and have excess resources, which creates a justification mentality. One of the sample participants remarked, "People have no respect for these big companies anymore; they hear about ABC president who made $2 million last year, and at the same time, they deny my grandmother for her visit to her general practitioner." Another remarked that it "becomes perfectly acceptable to rip off insurance companies when these high profits are publicized." It is important to note that not all insurance companies are involved with showcasing their profits. It seems that only a select few make this profitability highly public. Yet these few companies do create the mentality that all insurance companies are financially well-off. Of the fifteen members of the sample population, none of them reported that their respective companies were involved in the promotion of profits, but all of them mentioned two specific companies that seem to enjoy this publicity and furthermore contribute to the justification and rationalization of committing fraud.

While speaking at a medical conference in Florida several years ago, I performed an informal, anonymous survey of the session participants in regard to their opinion on fraud; it is worth noting that all eighty-five attendees were medical doctors. The survey revealed that almost 75 percent had at some point "made an alteration in the submission of an insurance claim in order to provide a more favorable result." This is not to imply by any means that the medical doctors at the conference were hardcore fraudsters, but it is interesting that a large majority replied in this manner.

We fully explored these results during the session and had an insightful dialogue about their experiences and opinion on insurance companies. The doctors remarked how they perceived insurance companies to be very process oriented and seem to focus to a greater extent on the billing and financial aspect of medical treatment than on the well-being of the patient. Several doctors, who had over fifty years' experience, provided stories of how the current state of insurance-to-doctor structure is such that the insurance companies drive treatment more than the doctors. One specific example came from a highly credentialed cardiologist who had a patient that he knew needed an EKG, but the doctor knew that it would only be covered by the insurance company if he used several alternative codes on the bill. The doctor had to use these codes in order for the patient to receive the much-needed EKG. The doctor's intention was not to up-code in order to receive higher bill payment, but simply to get this stressed patient the necessary medical care. To some, who subscribe to the zero-tolerance mentality, this would be considered fraud, but to others, it is simply an example of a knowledgeable doctor who has his patient's well-being in mind. One of the most significant aspects of this informal survey is that it tells us that the perception of fraud is acceptable among medical doctors, one of the most highly prestigious professional positions in our culture.

Fraud scammers often view committing fraud as a means to recoup premium payments; they feel they are owed payback for the premiums they have paid. Many offenders seem to quantify the premiums they have paid and think that this is money that is owed back to them. I recall reading a transcript from an interview of a convicted arsonist that was federally charged and prosecuted. He had insured several commercial buildings, and when he came into a dire financial state, he set them ablaze to collect the insurance settlement. In his interview, he commented on the exact premium he had paid over a five-year time frame and stated one of his main goals in the fraud was to recoup those premium payments. This mentality is quite common among fraudsters who believe that premium payments are a form of *loan* that they are entitled to redeem.

A compelling line of research, such as the study by Cartwright and Roach in 2016, shows how victimless crimes are easier to precipitate, as the offender does not feel that the crime has any negative impact, and, unfortunately, insurance fraud offenses, and most other white-collar crimes, are considered victimless crimes. There is significant monetary and humanitarian loss from these crimes, yet there is very rarely an actual *face* to fraud. We are well aware of the fact that the victim of fraud is the company, the policyholder, and the public, yet it is difficult to accurately identify a specific victim to showcase in these crimes.

We can also draw from other areas of criminality, such as homicide and theft, which tell us that when offenders do not perceive the actual impact of their actions, then they are more likely to commit the actual act. Criminological analysis of offenders in many other criminal areas has shown an alarming commonality that certain criminals precipitate crime because they have a significant personality flaw in that they cannot identify and *feel* the damaging effects of their actions. This lack of victim identification is significant, as it will make it easier for an offender to rationalize his or her actions in the commencement of crime. As we develop counter fraud efforts, we should publicize our efforts and the damaging impact of fraud to help create a perceived fraud victim. In one interview I performed of a convicted fraudster, he added merit to this argument and stated that he continued participating in fraud because it didn't make him *feel bad* like drug crimes he had previously engaged in. He expanded further and explained that when he dealt drugs, he could see the damaging effects of his deviant activity directly on the street, on a daily basis. Increased violence, addiction, poverty, and increased prostitution, and victimization were all the residual effects of drug dealing, which he could witness directly and visually. He stated when he turned to fraud, he never knew who he was damaging, and actually felt *better* about precipitating the crime as compared to his prior activity with drugs.

There also appears to be an alarming trend of generation Y's increased sense of entitlement, which undoubtedly contributes to the rationalization of this crime. During the past fifty years, we have seen a shift in the economy from manufacturing (blue-collar) jobs to professional (white-collar) occupations. The research has identified and categorized four main generations since that point in time: the silent generation, baby boomers, generation X, and the millennials.

The silent generation would be individuals born between 1928 and 1945 and would be children of World War II and characterized by loyalty, hard work, and high moral values. The baby boomers were born between 1946 and 1964 and are characterized by their loyalty to their careers and organizations. Generation X'ers would be those born between 1965 and 1981 and are known for their desire for work/life balance and maintaining a healthy environment for employment. Millennials, or generation Y, are born between 1982 and 2009 and are characterized as having a more narcissistic attitude than any of the members of the previous generations. This presents unique challenges, as this new group has entered the workforce with this newly found sense of entitlement different from all other previous generations.

The new sense of work ethic of the millennials was evident in several of the college courses I taught where a student asked what the minimum work threshold is that the student would need to complete in order to pass the course. Many references have been made to this generation by terming them the entitlement generation, characterized by the belief that they are owed certain benefits without real justification and they desire higher salaries, benefits, and more time off than the average worker. This sense of entitlement has drastic effects, as it serves to rationalize fraud activity; fraudsters can justify exaggerating claims, filing false reports, and attempting to increase settlement payout because they are *owed* this by society.

As we conclude our conversation of rationalization, ponder how your company or agency could develop strategies to help counter this rationalization mentality. We need to explore ways to reduce the justification component of the fraudulent act; we will explore specific counter fraud strategies in later chapters, so stay tuned.

Opportunity

One of the central themes of this book is to increase the awareness of vulnerabilities within a company or agency and develop strategies to reduce opportunity. This brings us to a discussion of the third and final component of the fraud triangle: opportunity. Opportunity refers to the specific ways that a fraudulent act can be committed. This is where the risk/reward scenario we spoke about earlier comes into application; the potential fraudster will see an opportunity and then weigh the respective risks and rewards in completing that action. Hopefully the risks will outweigh the reward, and fraud will be thwarted, but quite often, as indicated by our experiences, fraudsters take advantage of this opportunity and make a cognitive decision to participate in the fraud. One of the most important aspects in the assessment of the risk level by a fraudster is the potential for actually being caught; if there is a high probability of detection, then the risk factor will outweigh the reward, and fraud will be prevented. Opportunity reduction is the key to fraud fighting!

We must assume that we are all operating under somewhat restricted resources, in that we cannot simply implement a strategy based on unlimited staffing and financial resources. Our ability to reduce opportunities are highly dependent on the budget we are operating under and what areas we need to prioritize; fraudsters will recognize vulnerabilities and look to exploit those areas.

One of the main six themes that resulted from my dissertation study was focused on these resources. Most of the participants felt that competition over limited resources was the downfall of many fraud units, citing examples of instances where their units are passed over for resources in favor of other departments. One participant remarked: "Special Investigative Units are often the first to go when economics get tight. Like it or not, when you look at where we fit in with the company, we are always an expendable unit; companies ebb and flow, and from an overall perspective SIUs get cut because of this."

Other participants discussed the frustration they felt when seemingly basic fraud-fighting tools were cut from their structure, such as databases, technological items, and physical resources. These cuts in resources have a very significant ripple effect on our instances of fraud, as they create opportunity for potential offenders who will sense and exploit these weaknesses.

Another dissertation theme emerged that pertained to how insurance companies choose to pay and not fight. The sample felt that companies very often take the easy road and pay claims instead of fighting them. From one participant: "The problem is that insurance companies would rather pay out than fight a claim. It's a smart business decision in my opinion, but it sets a bad precedence. It sends a message to the criminals that they can go ahead and defraud them, as they will pay."

From another insightful sample member: "The insurance carriers are in a hurry to settle these claims; now with that mentality, you are going to create your own fraud. They make a business decision on certain cases to pay quickly; the claim adjusters are overburdened with claims coming in, and there is an urgency to pay the claim. I totally understand that, but this creates more fraud."

Several participants provided disturbing examples of how their companies often view fraud as simply lost business dollars and part of their normal business operations; they choose not to fight because they simply write it off as an expense. From a criminological perspective, as we recall from earlier chapters, this will convey increased vulnerability on behalf of the insurance companies and create the perception that more opportunities exist to commit fraud.

The fraud triangle is a widely used concept in many areas of fraud and can offer significant assistance with understanding the motivational factors of criminal behavior. The triangle helps to explain fraud behavior by providing us three areas to consider: pressure, rationalization, and opportunity. Pressure can come from personal or professional areas, such as job loss, financial difficulties, delinquent loans, tuition, or just simple debt, and greed. Pressure helps to explain what caused the crime to occur in the first place, and what made the offender consider fraud as an option. Rationalization is the second area of the triangle and focuses on how the offender will try to make him/herself feel *better* about committing fraud; fraudsters will minimize the fraud act by justification. The fraud act can be justified by payment of premiums, lack of victim identification, negative economy, entitlement theory, and many other factors. Opportunity is the final area of the triangle that we explored and focuses on how the actual fraud act can be committed. Fraudsters consider the risk/reward decision when assessing the opportunity or vulnerability within a company or agency and then act accordingly.

In the first three chapters of this book, an in-depth summation of the motivational factors that precipitate behavior was provided. The road map of this book is structured in a manner whereby we first discussed the fraud problem and established that this specific crime is a very significant problem that has wide reaching impact, impact of a financial and humanitarian nature. In the second chapter, we explored the mind of the fraudster and discussed this fascinating area of criminology, delving into the many psychologically based theories of motivation and behavior. Criminological theory is widely used and applied in many criminal circles, yet is highly underestimated and underapplied in fraud scenarios. We focused on four criminological theories that will offer the most useful application in our fraud world: rational choice, routine activities, deterrence, and strain theory. Investigating these four theories helped to illustrate their effectiveness and potential use in counter fraud applications.

Now that we have an intimate understanding of the motivational factors and psychology of the fraudster, let us apply these theories on a specific basis in our counter fraud efforts. In these next three chapters, we will use the concepts and principles discussed in the first three chapters and use them as a foundational structure to present and discuss specific counter fraud effort and strategies that can be implemented based on these psychological theories and concepts.

CHAPTER **4**

Counter Fraud Efforts

In Chapter 1, we established that fraud is a far-reaching problem with immense impact on many levels. We concluded that this specific crime is very worthy of further academic and research attention in order to provide us with the information to make accurate and effective counter fraud strategies. We understand that fraud offenders are more creative and sophisticated, yet we are operating with limited resources, which creates an interesting conundrum. We also came to a deep understanding of the financial impact of the crime, at a countrywide and personal level. In the second chapter, we explored criminological theory in detail and began with a general overview of the various criminal theories and their main tenets. We then delved into the mind of a fraudster and discussed the cognitive aspects of those that commit fraud, reviewing the potential application of criminal theories into preventative fraud strategies. The third chapter focused on the fraud triangle and how pressure, opportunity, and rationalization intermingle to create a fraudulent act. Examples of each of the three sections of the triangle were discussed, and specifics were extracted from my dissertation. The first three chapters of the book focused on the cognitive aspects of fraudsters, and the overall mind-set of the fraud offender. In the last three chapters, we will explore specific strategies for fraud prevention, addressing how criminal theory can be effectively applied in a counter fraud setting to develop highly strategic preventative policies and programs.

In this chapter, we will examine the detailed steps to develop counter fraud strategies, which include identifying risks and vulnerabilities, developing flags, integrating controls, monitoring and modification of controls, and analyzing measurement criteria. The vulnerability assessment will assist with identifying risks, the first step in this process. A full summation of the vulnerability assessment will be provided, which will include a detailed presentation of the Excel document entitled *Vulnerability Assessment*. The second step is to develop strong red flags as a result of the vulnerability assessment, flags that will assist keen fraud fighters of their high-risk areas. The third step is integrating controls, which is where we apply the strategies on an internal, external, vendor supported, or internally supported platform. The last crucial step in this process is the measurement of the program, and then monitoring and modifying as often as necessary.

Identifying Risks/Vulnerabilities

Studies reveal that almost 20 percent of companies have never performed a formal risk assessment, and of the 80 percent that have some familiarity with this process, most of them performed it inaccurately and ineffectively (Association of Certified Fraud Examiners 2016). The primary goal of this book is to assist fraud fighters with the development of highly focused and effective counter fraud strategies for immediate application in their respective organizations and agencies. It is recommended that the following multistep process be followed in order to assist with this endeavor:

1. Identify risks/vulnerabilities.
2. Develop workable red flags.
3. Integrate controls.
4. Monitor.
5. Measure.
6. Modify and reassess.

In order for a company or organization to develop effective counter fraud strategies, a structured risk/vulnerability assessment must be performed. It is critical for leaders and fraud fighters within a company to have a keen understanding of the specific areas of vulnerability in order to make proper decisions on fraud policies and procedures with laser-like, pinpoint accuracy. This is one of the largest areas of opportunity I have seen in the area of counter fraud policies: the general lack of data and information to support current or potential fraud programs and procedures. In the majority of companies and agencies that I have had the opportunity to work with over the past twenty-two years, I would estimate that over 75 percent of these organizations have counter fraud policies that are based on weak or moderately weak and potentially inaccurate information. As this weak foundation is the basis for their preventative fraud efforts, there is no doubt that many of these companies see very dismal fraud results from their efforts. This is not to imply that these companies employ individuals that are lacking in leadership and managerial skills; it seems to be more of a misunderstanding on how information and data are developed.

This is where there is an incredible academic opportunity: the insertion and development of fraud data and information using a highly rigorous academic approach. Prior to entering into academics over ten years ago, I was not fully aware of the structure, process, and procedure for publishing an article. I assumed that as long as an article was in some sort of publication, no matter the level, it meant the information was accurate, credible, and worthy of application. Not to say I was naïve, but I assumed that if the article somehow made it into a printed copy of a magazine, newspaper, or other medium, then it had been filtered through many layers of approval before printing. In reality, I realized later that this is often not correct. This is not to insinuate in any manner that those that are publishing articles are all ill-informed and not worthy to author; it is simply to make mention of the publishing process so that those in our industry can be more guarded with the specific application of information. In effect, most of the articles that are written and distributed are based on opinion and not fact, which makes them highly inaccurate.

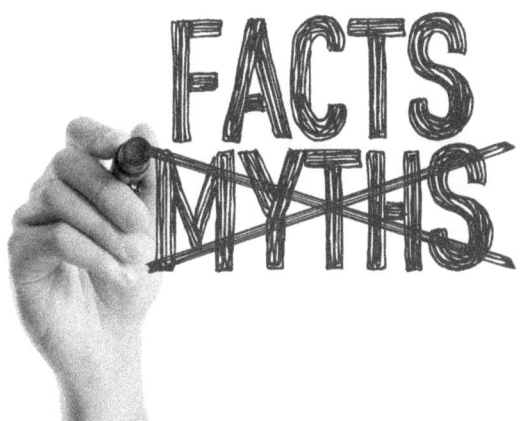

There is a misunderstanding in nonacademic circles over the meaning of using the term *I am published,* but in academic circles, to be *published* means that you have written an article that has been peer reviewed by others in academics. The term *peer review* is the most important key to publishing an article and implies that other academics, mainly PhDs, have rigorously, and often relentlessly, reviewed, assessed, and provided in-depth feedback on an article.

Harvard offers ("Evaluating Journal Articles," n.d.):

> When searching for journal articles, it's best to find articles that have been vetted by scholars in the field. Editors of refereed or peer-reviewed journals send prospective manuscripts to scholars who specialize in the topics covered, and these scholars critique the manuscripts without knowing the identity of the author. If an author has made claims that are unsubstantiated or considered problematic by his or her peers, the manuscript may not be published; if, on the other hand, the manuscript is deemed rigorous in its argument, it will be published. The review process is meticulous and dispassionate (since the reviewers don't know whose manuscript they're reading, they can't play favorites). By contrast, journals that are not refereed publish manuscripts that have been reviewed only by an editor or editorial collective, and these manuscripts are not reviewed anonymously. In print sources, information about the editorial process is usually available near the front matter of the journal; in electronic sources, clicking on the journal title will usually lead to a page outlining the editorial procedures.

Additionally, the John Jay College of Criminal Justice defines peer review as ("What is a Peer Reviewed Article?" n.d.):

> In academic publishing, the goal of peer review is to assess the quality of articles submitted for publication in a scholarly journal. Before an article is deemed appropriate to be published in a peer-reviewed journal, it must undergo the following process:

- The author of the article must submit it to the journal editor who forwards the article to experts in the field. Because the reviewers specialize in the same scholarly area as the author, they are considered the author's peers (hence "peer review").
- These impartial reviewers are charged with carefully evaluating the quality of the submitted manuscript.
- The peer reviewers check the manuscript for accuracy and assess the validity of the research methodology and procedures.
- If appropriate, they suggest revisions. If they find the article lacking in scholarly validity and rigor, they reject it.
- Because a peer-reviewed journal will not publish articles that fail to meet the standards established for a given discipline, peer-reviewed articles that are accepted for publication exemplify the best research practices in a field.

Those that operate in the academic arena are intimately familiar with this peer-reviewed process and conduct research, investigate, and write articles accordingly. The goal of providing this summation of peer-reviewed articles and academic writing protocol is not to create widespread panic and paranoia in regard to the state of information that is currently in circulation in our industry, but the goal is to simply create an awareness of the academic process, a process that was unfamiliar to me prior to becoming involved in academics. An awareness and familiarity of this will help to assess the relative efficacy of the articles, data, and information that is being circulated in order to pick and choose the information that is the most credible and relevant to your particular fraud challenges.

An example of a peer-reviewed publication would be a completed dissertation or a peer-reviewed journal. The completion of my dissertation was a four-year process whereby three committee members, who were all well-established and published PhDs, provided hundreds of recommendations, which resulted in countless iterations. Additionally, publishing an article in an academic journal requires review, approval, and feedback dozens of times from the editors of the journal, who, similar to dissertation mentors, are well-established and published academics. This rigorous process is to ensure that the information provided to the public is accurate and worthy of application.

One of my biggest fears when I attend conferences is when I listen to a presentation that is based on opinion, and I witness the attendees of the session vigorously taking notes on the information provided. I was speaking at a conference in Austria this past year and decided to listen in on a neighboring session where the topic was application and underwriting fraud. The session first focused on trends in application fraud and then on methods to avoid this type of misrepresentation. As he presented his PowerPoint, the information he mentioned on trends was entirely incorrect; as a matter of fact, I read an article (peer-reviewed, of course) earlier in the week that provided data that was the exact opposite of what he was discussing! I watched the attendees taking notes and cell phone photos of his PowerPoint with the intention of communicating and potentially applying this information within their respective companies.

This is also true of many articles that are readily distributed to fraud fighters, articles that discuss fraud strategies, data, and trends, and are authored by individuals that may not be qualified to offer insights. My recommendation is to read articles and attend as many conferences as your time permits, but be extremely guarded with the origin of the data and how you will apply it within your organization. Question the source of the data, and make sure it is based on credible, reliable research.

The core message to be delivered from our discussion of academic writing is to make sound decisions based on sound information. And the term *information* is synonymous with data. Therefore, when one is assessing corporate vulnerabilities, we must apply what we have learned and base it on solid, credible data. The very first step in assessing risk is to obtain information and data that is current and reliable, which leads us to our first conundrum, the measurement of fraud.

Based on the research currently in circulation, my specific research as faculty for Colorado State University-Global Campus, and consultation with dozens of companies, one of the most significant challenges facing our industry is the relative lack of awareness of the fraud problem. That is, there is no universal or consistent measurement of fraud, which makes it difficult to accurately measure and report on the problem. As an industry, on a global level, we do not have one specific agency that collects fraud data; each insurance company, state or federal agency, or any other organization must devise its own reporting protocol.

There are many companies that measure fraud by single occurrence; some measure by prosecution, some by exposure, and others by dollars collected, and the list continues. In an informal experiment I performed, I asked four auditors to review the same exact claim files for potential fraud. The reviewers were consistent with their percentage of identification; each identified 7 to 12 percent as suspicious, yet not one of the same claims was labeled as fraudulent by all four auditors! In other words, they all picked different claims as having fraudulent indicators. There also appears to be an interesting ripple effect when considering the relative efforts of fraud identification. If a company invests very little time and effort into detecting and attempting to quantify fraud, then little fraud will be uncovered, providing a false sense of the problem. If preventative strategies are based on a meager detection approach, then the company will undoubtedly have a false sense of security due to low detection results, which could be quite contrary to its true fraud exposure. A self-developed approach toward fraud measurement will be introduced in the last chapter, an approach that serves to create consistency in measurement.

Results from my dissertation study interviews also revealed interesting information in relation to this topic. A second research question posed to the sample asked for their opinion on what barriers and challenges are reported by insurance fraud investigators in helping fight insurance fraud. The thematic analysis of their responses revealed the following six themes:

1. No political or judicial support.
2. Financial barriers.
3. Fraud is difficult to quantify.
4. Fraud is a social problem.
5. Claims staff changes.
6. Companies are their own worst enemy.

Almost all fifteen study participants felt that one of the largest barriers to effective fraud prevention was the fact that fraud is difficult to quantify. Each participant revealed his or her specific company's measurement process, and all fifteen companies represented in the sample used entirely different approaches in how they measure fraud.

One sample participant stated, "Without these impact generating statistics, fraud will never get the attention that it deserves."

Another stated, "There is absolutely no rhyme or reason to fraud reporting; if you were to ask 10 different fraud fighters, you would get

10 different answers on how they measure fraud. Fraud is not black and white. If you run a red light, you know what the violation is; you know the cost and what the charge will be. Fraud is not like that; you never know what you're dealing with, what the costs are."

Many members of the sample discussed that the lack of centralized reporting is a major problem as well, citing specific examples of disparity in reporting. Many insurance carriers have to report fraudulent claims to their respective fraud bureaus, which are responsible to maintain records and statistics on fraud crimes. However, as reported by the sample, many of these fraud bureaus are operating under limited resources and as such do not have the time or resources to accurately maintain these fraud statistics.

Additional inconsistencies exist with the definition of simple fraud terms. We all freely use the terms *insurance fraud*, and quite frequently we hear colleagues discussing their relative insurance fraud rate. Yet in our industry there are dozens of different terms used to define specific types of fraud. Some common definitions are as follows:

- Hard fraud: large-scale fraud cases usually involving criminal prosecution and large sums of money. Organized crime rings are common in this category.
- Soft fraud: the most common type of fraud, which involves the misrepresentation of facts.
- Planned fraud: this could be committed by a group or an individual and involves a totally fabricated loss such as a staged accident.
- Opportunistic fraud: This type of fraud exists when an individual takes advantage of the opportunity of the legitimate claim. There is usually no forethought.
- Exaggerated fraud: when an individual suffers a legitimate loss and adds additional items or damage.
- Medical fraud: also known as medical provider fraud and exists when medical facilities overbill or bill for services not rendered to increase reimbursement.

- Underwriting fraud, also known as application fraud: intentional misrepresentation of information at the policy inception stage in order to obtain a reduced premium.

As we can see, many of these terms overlap and have similar meaning, yet we use them regularly as part of our daily professional vocabulary. Each could be measured and tracked differently, which again creates an inconsistent picture of the fraud problem.

To illustrate this inconsistency, the following are several examples of data and measurement disparity based on my consultations. Company A measured fraud based on dollars saved; that is, its annual fraud result was measured by the total number of dollars saved as a result of its fraud investigations. When further inquiry into this monetary amount was requested, it was discovered that the measurement appeared to be based not on the amount that the fraud investigation actually saved or recovered, but on the reserve of the claim file at the time it was transferred into the fraud unit. If claim 123 was transferred into the fraud investigations unit with a reserve of $50,000 aggregate for all exposures, then this fraud department took a savings of $50,000 for this file, regardless of the outcome of the actual investigation. Another company (B) would measure its fraud rate by the percentage of cases where fraud was proved, that is, claims that it can deny. Company C would measure its fraud rate by the number of cases that enter into the unit, regardless of whether fraud is proven, or the claim simply sent back to the claims department for payment. As we can see, there are many different methods to measure and track fraud, and each company and agency seems to have a different method for this process, a method that creates inconsistent data and a false sense of the true nature of the problem. Two large companies based internationally add a unique perspective to this topic. Both made a conscious decision to refrain from measuring and sharing their fraud rate and accompanied savings, as they feel it gives the appearance that they take pride in denying their valued policyholders, a message they do not desire to disseminate.

The point of discussing measurement inconsistency in fraud data is not to create a doom and gloom perspective, it is to bring awareness to the fact that one of the major problems we have facing us in the fight against fraud is the lack of a universal definition. This lack of common ground makes it difficult for us to accurately communicate our problem to others within the industry and also to legislators and the political arena. Thus, one of the specific strategies we can start to ponder as fraud fighters is a way to create a benchmark measurement of fraud by creating common definitions and a formula for fraud identification. The challenge here also lies with the release of sensitive, confidential corporate information across companies; many organizations are hesitant to release specific information as they fear it may violate corporate policy. An in-depth discussion of fraud measurement will occur in the last chapter, where I will make recommendations and introduce ideas on more consistent methodologies to measure and collect data on fraud within our companies. This consistent approach is very important if we are to truly understand the nature and scope of the problem we are facing.

As we start to discuss specific methods to perform the vulnerability assessment, the objective of all carriers and agencies should be to start with reliable data. Data that is reliable and credible will provide a strong foundation for highly focused and effective prevention strategies. Utilize an academic approach as we discussed earlier and question the origin and collection procedure of the data that you are provided in order to ensure its accuracy. Once you have data, the fun begins!

Assessing the risk and vulnerabilities within an organization should be a formalized, structured process that occurs at least once a year. A full vulnerability assessment could take place annually and a check-in every quarter in order to monitor and modify according to trending data. The Vulnerability Assessment Worksheet as shown on page 76 is a formalized approach to conducting a risk assessment and can be used as a tool for immediate application. This assessment can be initiated internally or with the assistance of an external consultant.

One of the main messages of this book is to communicate to carriers and agencies the need to develop some sort of counter fraud strategy, any strategy! I have been exposed to too many companies that either do not have a system in place or have a system that is informal and unstructured; both of these scenarios have drastic results. As we delve into the specifics of the vulnerability assessment worksheet, it is important to keep in mind that this worksheet can be completed in a highly simplified fashion by one or two individuals over coffee, or extremely structured, which would include robust data, an external consultant, and a team of colleagues. The latter, more in-depth analysis is the recommended method for completion, as the time and resource investment is well worth the benefits, but, again, if time and financial resources are limited, any strategy is better than none.

Let us now turn our attention to the specific columns on the assessment worksheet and discuss their relevance and completion process.

Vulnerability Assessment Worksheet

Vulnerabilities	Likelihood 1–10	Significance 1–10	Department	Existing Fraud Controls	Fraud Controls Effectiveness 1–10

Assessing Vulnerabilities

The first step in conducting a vulnerability assessment is to identify the risk and vulnerable areas within a company or agency. It is important to keep criminological theories and concepts in mind while investigating this very first area of the process. As we recall from our in-depth discussions in chapter 2 on criminological theories, and chapter 3 on the fraud triangle, effective fraud fighting should focus on reducing opportunity. Fraud offenders look for opportunity and make conscious decisions to commit a fraudulent act; therefore our vulnerabilities are translated into opportunities for fraudsters. Vulnerabilities can come from many different sources within a company, and it is important to look in all areas within the organization. Often, during an initial training session, I will ask the attendees to brainstorm and write a list of every single area of fraud that could exist within their company, even if it has not been an issue in the past. This unlimited structure creates a starting point for the vulnerability assessment. Below is a list of potential areas and subareas to focus on when conducting the vulnerability step.

1. Sales
    a. Internal agents. Is there any system of monitoring agent and sales activity? One of the largest cases I investigated while in SIU involved an internal sales agent who misrepresented commercial policy information for personal gain. This was a very involved case and involved multiple state and federal agencies, but it would have been easily discovered with any form of simple software or internal detection system.
    b. External agents. Is there any method to monitor activity? A current trend in Europe, and moving to the United States, involves ghost brokers—that is, brokers who sell policyholders entirely fake policies.
    c. Agency staff. How much access do they have to personal information?

d. Sales notes. Is there access to the sales notes? There is a plethora of information that can be gained from the sales notes. There are many instances where an insured will call his or her trusted sales agent/administrative assistant in order to check on coverage. The policyholder will ask a hypothetical question to the agent and then remarkably suffer a legitimate loss only days later that is relevant to the earlier inquiry.
   e. Claims reporting process. Are the sales staff given a monetary authority to settle claims? One carrier I worked with allowed its agents to settle claims under $500 directly from its offices with no oversight.
   f. Commission structure. One large carrier based its agents' commission on sales minus number of claims reported from their book of business. It was therefore monetarily advantageous for agents to report as few claims as possible, which led to delayed reporting and an unknown number of missed fraudulent claims.
   g. Training. Does the fraud department offer any type of training to the sales staff to increase awareness of the fraud problem and also communicate specific fraud trends and patterns?
2. Marketing
   a. Public outreach campaigns. What is the public's opinion of your company's tolerance to fraud? In the last chapter, we will discuss public opinion in depth, but we already briefly discussed how the public generally accepts fraud. One carrier I worked with prefers to showcase a strong presence of zero tolerance to fraud by communicating this very clearly in many policy documents and also by sharing stories of successful fraud prosecutions. Another carrier I consulted with had the exact opposite corporate opinion and culture and felt that showcasing fraud fighting is very damaging to its public opinion of being a policyholder-friendly company.

3. Underwriting
   a. Staff changes—increase or decrease? Reduction in staffing in underwriting is a common and very significant vulnerability.
   b. Any changes in application process? Look for reduced standards such as Internet policies, which are very vulnerable areas and make it easier to misrepresent policy information.
   c. Is there a policy verification process? In many carriers, the SIU or investigative units would be tasked with verifying policy information on certain high-exposure policies, such as commercial buildings and high-value residential homes. This is a highly vulnerable area if there is no policy verification process in practice.
   d. New local/state/federal rules or regulations. It is important to keep abreast of any underwriting standards that could result in insuring more high-risk policyholders.
   e. Residency verification. Misrepresentation of one's residency is one of the top three trends in premium fraud. Do we have a method to verify this, such as asking for a work ID, utility bill, or conducting data searches comparing policy zip code to the insured's cell phone area code.
   f. Vehicle mileage verification. This is a second trending area of premium fraud in auto carriers: insured's who understate the mileage usage of their vehicles. Can this be verified through databases such as CARFAX or highway toll pass programs such as New York's EZ Pass?
   g. Commercial use verification. This is a third trending area of premium fraud and involves an insured using a vehicle for commercial purposes when it is rated for personal use. A common commercial scam involves a policyholder who uses an auto for taxi or livery purposes and is only rated by its carrier for personal use. This can often be verified by Google maps, social media, and interviews with the passengers (if a claim is filed), police report details, and a covert inspection of the insured's vehicle.

4. Loss report process
   a. Who are the loss report takers? Are they experienced in potential fraud red flags? Statistics show that the loss report is a critical point of a claim; if the insured is not treated well on this first call, then they will often feel justified at committing fraud, as discussed earlier in this book. If insureds feel that the insurance company is trying to get one over on them at this early stage, this often creates a justification mentality that breeds fraudulent activity.
   b. Notes from loss report. Crucial information can be gleaned from the notes taken at the time of report. In cases of opportunistic fraud, this is often the moment where the insured may realize he/she has an opportunity for additional damage and/or to inflate the claim. Do we have access to these?
5. Claims
   a. Fast-track units. How many cases are going to these units and what is the system to route them there? Are we missing fraud cases because they are under our fast-track threshold?
   b. Processors. Similar to fast-track units, these employees handle routine tasks in the claim process. Are they trained to look for fraudulent claims?
   c. Optimization. Is there any opportunity to optimize, or even automate, the referral process to the fraud unit?
   d. Claim representative experience. Does the staff in the claims units have the experience and guidance to identify fraudulent claims?
6. SIU-fraud units
   a. Reporting process. How do claims get to SIU, and are there any obstacles to this process? Is this a manual or an automated procedure, and are we missing opportunities? Oftentimes the fraud referral process is tedious and time-consuming; many claim representatives refrain from transferring files over simply because of this workflow issue.

b. Intelligence. Are we getting all of the valuable external intelligence we need to operate effectively? What are the barriers to gaining this information? This would include up-to-date information on rings, medical provider cases, and current trends that are occurring in a carrier's coverage area.
   c. Resources. Do we have all of the necessary resources to productively handle fraud cases? Do our investigators have database access to verify certain information, which could reduce field time?
7. Subrogation (recovery) units. Are we ensuring that these units are aware of the potential fraud that could be identified at this late stage in the claims process?
8. Vendors and outsourced partners
   a. Kickbacks? In any situation when a company utilizes outsourced partners such as investigative firms, legal defense firms, accident reconstruction companies, medical review and exam companies, and field appraisers, someone internally must manage this vendor relationship. In this situation, there is always the possibility of a kickback or referral scheme.
   b. Oversight of invoicing? In addition, when vendors are used within an organization, an internal employee must also issue payment; this is also an opportunity for internal theft.
   c. Communication with vendor. The communication channel must also be assessed when looking for vulnerabilities. If a vendor is assigned to a case, and the vendor uncovers new information, how is this information relayed back to the claims/investigative unit? Are we missing fraud flags in this process?
9. Field appraisers
   a. Communication. Is there open communication with the main file handler or investigator if new information develops?

b. Training. Does the fraud unit offer training to the appraisers and other staff in order to communicate trends and patterns of fraud that they should be looking for?
10. Internal employees
    a. Threshold for claims payments. This is becoming an increasing concern of many carriers and agencies. Software programs, both external and internal, can be utilized to filter payments and look for trends that would indicate an internal employee is issuing payments to himself or herself. Oftentimes, the internal employee will make multiple payments below his or her settlement threshold in order to avoid detection.
    b. Computer access. An assessment of who has access to certain programs and corporate information should also be performed. There should be a strong system of checks and balances whereby each task (such as a claim payment) can be checked and verified by another employee (usually the manager).
11. External threats
    a. Weather patterns—earthquakes, hurricanes, etc.
    b. Competitive threats. Are others in the market taking a different (more or less aggressive) approach to fraud fighting, and is this making the company more vulnerable?
    c. Economic trends. Is there a poor economic environment that would cause an uptick in opportunistic fraud?
    d. Other trends? As discussed earlier, certain auto loan companies have reduced their financial standards to offer loans to those that would normally be unqualified. This has resulted in more insureds driving vehicles that they cannot afford and are therefore more likely to attempt or to ponder an owner give-up scheme.

Questions to ponder as you investigate vulnerabilities are as follows:

1. How might a fraudster exploit weaknesses in the system of controls? It is important to address all areas above as potential vulnerabilities.
2. In what area does the company or agency suffer losses (vulnerabilities)? This information can often be obtained by annual records, management reports, or using internal or external software capabilities.
3. Where have we suffered losses in the past? As in #2 above, this can come from reports and other numerical sources, but also can come directly from key individuals working within any of the areas discussed earlier.
4. What are the current market/external trends and where will we potentially suffer losses in the future?
5. What external factors should the company be most concerned with?
6. What internal factors should the company be most concerned with?
7. What is the state of our financial and employee resources, and will this change?
8. Has the corporate/leadership philosophy changed, and how will this impact fraud strategies?
9. What are our current fraud red flags, and are they accurate? When was the last time we modified red flags to reflect current trends?
10. Are we writing the same lines of business or have we expanded into new areas? Will the fraud department be consulted with before and during this process? In one large company, the leadership focus was on premium written, and as such decided to focus on a state in the Northeast to increase business activity. What the sales and leadership staff was not aware of was how this area was at extremely high risk for fraud. Ironically, a strong

insurance competitor withdrew from this state a year prior because it experienced significant losses due to fraud.
11. Are we writing in the same states, or have we expanded our territory?
12. Has anything changed in our sales structure? Are we using exclusive agents, multicarrier agencies, the Internet, and so on, to market our insurance products?
13. Have we experienced any downsizing or other threat that could cause dissatisfied internal employees and increase the occurrence of internal theft?
14. Are we outsourcing any tasks, and who is managing these vendor relationships?
15. Have there been any changes to the underwriting structure within our organization, and are there any new application/underwriting regulations to which we must comply?
16. Are we continuing to monitor and update our data analytics or software system to tackle new threats and scams?
17. What information are we using to make our counter fraud policies; is it academic and current?

These areas and questions are not all-inclusive; many more can, and should, be explored within an organization. To glean the most value from this assessment, it should ideally be a team approach; those from all pertinent departments should be consulted with and provided the opportunity to share data and information. The most effective method that I recommend is to create a formal committee or task force that contains members from all units within an organization. The vulnerability assessment procedure starts with identifying the appropriate risk and then formally assessing and prioritizing its relative vulnerability.

The results from internal audits can also provide very useful data for assessing vulnerabilities. As discussed earlier, it is common in our industry to assume that 10 percent of all claims are suspicious. However, in one audit I reviewed, it revealed that the percentages of fraud buildup was significantly higher in no-fault states—Florida 31 percent, New York 24 percent, Massachusetts 22 percent, and Minnesota 22 percent. This would be incredibly useful information to integrate into our vulnerability assessment, as it would help us focus on the areas where we can have the most impact (PricewaterhouseCoopers 2016).

Completing the Vulnerability Assessment Document

When completing the vulnerability assessment document, the first column, entitled Vulnerabilities, can include any area of risk within a company as listed above. Once risk areas have been identified, the worksheet should be utilized immediately, which will assist with organizing and developing more focused counter fraud strategies; it should be considered a working document and not one where the goal is to have a final, completed copy that is not manipulated and modified.

The ideal method to hone in on vulnerable areas is to start with broad terms and then narrow one's focus to subsequently create subcategories for additional analysis. For example, in Insurance Company A, there was an unfortunate downsizing within an underwriting department, resulting in a reduction of underwriters from three to one. Therefore, underwriting would be listed as a first-priority vulnerability. The second step is to delve deeper into this risk area (underwriting) and develop more focused subcategories for analysis. We understand that as a result of this reduction in underwriting staffing, there will be fewer employees to review new policies, and as such there is a higher probability that there will be more cases of policy misrepresentation in the areas of residency and mileage. As these are both areas of concern for fraud units, these two subcategories should be listed as in the example below. Underwriting is the first priority, and then policy-misrep—residency, and policy misrep—mileage are the two associated subcategories. It is important to again reiterate to focus broadly on the first-priority items and then list as many subcategories as applicable; they can always be removed later. The subcategories are what we will be scoring in the worksheet, as they are more focused than the first-priority items. The completed column will look like this:

Vulnerabilities	Likelihood 1–10	Significance 1–10	Department	Existing Fraud Controls	Fraud Controls Effectiveness 1–10
1) Underwriting— downsized from 3 to 1					
a) Policy misrep— residency					
b) Policy misrep— mileage					

Now let's also assume that our vulnerability investigation revealed that our company suffered significant losses in prior years from the increasing prevalence of medical provider fraud. It was discovered that one of the metropolitan areas that we write business in has seen a very significant increase in medical billing fraud. We have decided that this will create a significant risk for the company in the upcoming year and therefore should be included in the worksheet. We will use medical billing fraud as a level 1 priority item and then start to develop subcategories. When we delved further into this type of fraud and reviewed the managerial and statistical reports secured from our analyst, we see that we suffered the most losses in cases where unscrupulous medical providers submit bills for services not rendered and up-coding scams. We can also estimate that our projected losses in medical fraud totaled around $5 million last year. Thus the worksheet will now look like this:

Vulnerabilities	Likelihood 1–10	Significance 1–10	Department	Existing Fraud Controls	Fraud Controls Effectiveness 1–10
1) Underwriting— downsized from 3 to 1					
a) Policy misrep— residency					
b) Policy misrep— mileage					
2) Medical provider fraud					
a) Billing for services not rendered					
b) Up-coding					

Now we can move into the next column of the worksheet—Likelihood, which is measured by a 1–10 scale. In this heading, the main question to ponder is to what extent a specific vulnerability is likely to occur, 1 being least likely and 10 being most likely. It is not critical to focus on the exact value, as there is no mathematical formula to assist with this computation; it is designed to be a generalized value. The most valuable aspect of the worksheet lies in its ability to create a visual snapshot of the vulnerable areas and helps to focus efforts. Once the worksheet is completed in its entirety, it is possible to view all columns and vulnerabilities and reveal which areas are worthy of attention. Thus, it is not critical that the values be specific to the .001, but only accurate enough to assist with the comparison between vulnerabilities.

In our case, we have two vulnerabilities that we are going to compare

PSYCHOLOGY OF FRAUD

and contrast: underwriting and medical provider fraud. We will now walk through each priority vulnerability and subcategory in the first (A) column of the worksheet and only score the subcategories, as they are more focused. The likelihood of policyholders misrepresenting mileage and residency information is a 10, as both of these scams will likely increase as we now have fewer underwriting staff to help identify these occurrences. We will accordingly insert a 10 into both columns B3 and B4. As we focus on medical provider fraud, the second priority item, we know from our statistical review and investigation that the likelihood of increases in providers billing for services not rendered and up-coding is also very high; thus we insert a 10 in B7 and B8. The completed sheet will now look like this:

Vulnerabilities	Likelihood 1–10	Significance 1–10	Department	Existing Fraud Controls	Fraud Controls Effectiveness 1–10
1) Underwriting— downsized from 3 to 1					
a) Policy misrep— residency	10				
b) Policy misrep— mileage	10				
2) Medical provider fraud					
a) Billing for services not rendered	10				
b) Up-coding	10				

The thought may have crossed your mind as to if there will ever be a value under a 10 in the B-Likelihood column. After all, if the vulnerability is worthy to make the worksheet, than there is a strong (10) likelihood that it should occur, correct? Not necessarily; there could be circumstances where we have identified the vulnerability but then realized that the likelihood is not very strong. For example, a scenario could occur whereby we have heard that there may be a reduction in underwriting staff from three to one, but this has not occurred and has not been verified. In this instance, we would want to list it as a vulnerability along with the subcategories, but the likelihood of occurrence may be only a 5 or 1. Again, it is important to list all potential vulnerabilities, as this will create a global snapshot and help to focus efforts.

As we move into column C-Significance, it is important to have an understanding of the potential exposure of the vulnerability. Oftentimes carriers will use management or annual reports, or utilize the assistance of internal or external IT employees to run reports to help assess exposure. Again, if the number is not precise, this is not a cause for concern; use a generalization based on the information available. In our scenario, we will start with C2 and move downward to C8. Because we are a carrier that writes a significant amount of subpar, or assigned, risk policies, we can estimate that the underwriting significance will be high. Our data analysis into policy misrep—residency, and policy misrep-mileage has revealed that there is significantly more exposure to residency fraud over mileage fraud. Therefore policy misrep—residency earns a 9 (C3), and policy misrep—mileage earns a 5 (C4). As we move to medical provider fraud, our analysis of billing for services not rendered and up-coding reveals very alarming statistics significantly higher in exposure than underwriting, thus billing for services not rendered and up-coding will earn 10s accordingly. The updated worksheet will now look like this:

PSYCHOLOGY OF FRAUD

Vulnerabilities	Likelihood 1–10	Significance 1–10	Department	Existing Fraud Controls	Fraud Controls Effectiveness 1–10
1) Underwriting—downsized from 3 to 1					
a) Policy misrep—residency	10	9			
b) Policy misrep—mileage	10	5			
2) Medical provider fraud					
a) Billing for services not rendered	10	10			
b) Up-coding	10	10			

The Department column is designed to show a visualization of the departments that a specific vulnerability affects. This could be S-Sales, U-Underwriting, C-Claims, S-SIU, or others. An organization can develop these symbols as they deem appropriate. The purpose of this column is to show how impactful a specific vulnerability is; obviously the more departments that a vulnerability affects, the more significant it may be within the company. In addition, it is also possible to see which departments are the most vulnerable and can also be used to sort and filter by department. In our case scenario, we know that (1) underwriting will affect the underwriting department directly, but there will also be associated departments affected, such as claims and SIU, which will face additional workload as a result of the reduction in staff at the underwriting level. In (2) medical provider fraud, the two departments affected will be claims and SIU, so we would enter C and S accordingly. The sheet will now look like this:

Vulnerabilities	Likelihood 1–10	Significance 1–10	Department	Existing Fraud Controls	Fraud Controls Effectiveness 1–10
1) Underwriting—downsized from 3 to 1					
a) Policy misrep—residency	10	9	U,C,S		
b) Policy misrep—mileage	10	5	U,C,S		
2) Medical provider fraud					
a) Billing for services not rendered	10	10	C,S		
b) Up-coding	10	10	C,S		

Existing fraud controls. For each vulnerability listed, are any fraud controls currently utilized? This section of the worksheet requires that an assessment be performed to address the specific controls for each risk identified. If there are no controls currently being used, *none* will be inserted in this column. This column is more free-form based, and open text is encouraged. In our underwriting vulnerability example, the only form of control we have in place is the underwriting staff identifying and filtering residency and mileage issues; therefore, we will insert text such as *manual* or *employee only*. If our underwriting department utilizes some form of internal or external software or database system to assist with filtering, then we can use such terms as *software, analytics, Excel*, and so on. In our medical provider fraud vulnerability, we have a software program in place to help us identify fraudulent billing; however, this system is an internal Excel system and is very basic in its capabilities. We would therefore use such terms as *internal software* or *internal system*. If we had an external system, we could use *external software*, or *external predictive analytical system*. As mentioned, depending on the vulnerability, there are limitless possibilities for text in this column, including *SIU, management oversight, external intelligence*, and so forth. Our completed column will now look like this:

Vulnerabilities	Likelihood 1–10	Significance 1–10	Department	Existing Fraud Controls	Fraud Controls Effectiveness 1–10
1) Underwriting—downsized from 3 to 1					
a) Policy misrep—residency	10	9	U,C,S	Manual	
b) Policy misrep—mileage	10	5	U,C,S	Manual	
2) Medical provider fraud					
a) Billing for services not rendered	10	10	C,S	Internal Software System	
b) Up-coding	10	10	C,S	Internal Software System	

The final section of the worksheet focuses on the effectiveness of the existing fraud control currently in place, on a 1 to 10 scale, 1 being least effective and 10 being the most effective. As with several of the other columns, the exact figure to the .001 is not necessary; as long as there is a general understanding of the effectiveness, then inserting a value in this field should not be difficult. This value can be based on reports and databases or on general observation. In our example, our underwriting vulnerability is currently utilizing a manual system of fraud control; coupled with the downsizing of staff by two employees, we can estimate that the fraud controls effectiveness would be very low, at a 1 or 2 for both residency and mileage. Regarding medical provider fraud, we understand that we do have analytics to help us filter out fraudulent billing; however, as it is an internal system and very basic, we can estimate that the effectiveness is approximately 5 or 6. Therefore our completed vulnerability assessment worksheet will look like this:

Vulnerabilities	Likelihood 1–10	Significance 1–10	Department	Existing Fraud Controls	Fraud Controls Effectiveness 1–10
1) Underwriting— downsized from 3 to 1					
a) Policy misrep— residency	10	9	U,C,S	Manual	1
b) Policy misrep— mileage	10	5	U,C,S	Manual	2
2) Medical provider fraud					

| a) Billing for services not rendered | 10 | 10 | C,S | Internal Software System | 5 |
| b) Up-coding | 10 | 10 | C,S | Internal Software System | 6 |

The completed worksheet is meant to create a global visualization of the state of fraud in our particular organization; it allows us to identify areas of opportunity for increased focus. We can develop highly focused counter fraud programs based on the results of this assessment, as we can identify areas that we should focus on based on our resources. The vulnerability assessment is not designed to be extremely difficult to complete, requiring volumes of data and reports and dozens of meetings and committees. Furthermore, each column is meant to be intuitive and also is designed to be completed based on minimal information and research. As I have mentioned hundreds of times in speeches, consultations, meetings, and one-on-one conversations, companies need to have some system in place—anything!

This worksheet should create that highly needed visual snapshot of the state of fraud within an organization; viewing overall vulnerabilities in this broad manner will assist in identifying outliers and help sort data and prioritize fraud areas. Visualization is a very important aspect of effective data interpretation; it helps to show the data and relative connections between information. As stated, this vulnerability worksheet is intentionally designed for its ease of use and the ability to arrive at column values without time-consuming statistical formulas or technical algorithms to complete. If other highly technical statistical analysis is desired, there are many other options to be considered, but they will require significantly more time and effort to be invested. Of course, many software programs are available to assist with this, such as SPSS; however, these systems require a time and financial investment in order for one to become proficient.

For example, once basic numerical data is collected and quantitative information is compiled, many simple statistical analyses can be performed. One of the easiest statistical methods to use is to simply summarize the data in a manner so it may be grouped and organized, and this can be accomplished by using various forms of graphs, such as bar graphs, line charts, pie charts, and histograms. Using these methods, one can group and filter data into different categories for visual interpretation. A second statistical method is to calculate the mean, median, and mode of the data in order to analyze the average values of the quantitative information. Third, one could also calculate the range, standard deviation, and variance of the data, which will reveal the overall spread from the core values.

A more advanced technique would be performing a linear regression analysis in order to compare variables. Linear regression is a foundational form of predictive analysis and is the most commonly used form of regression. Basic linear regression describes data and the relationship between one dependent variable and one or more independent variables and is used specifically to (1) analyze the correlation of data, (2) forecast the effect, and (3) forecast the trend. For example, linear regression can be used to focus on insurance claims and factors that could most accurately predict fraudulent activity. The independent variables could be (a) number of years with the insurance company, (b) claims history, (c) claims per year, (d) high-risk policy, (e) new business, and (f) police involvement. The dependent variable would be fraudulent claims and non-fraudulent claims. A linear regression would then reveal the most significant correlation, .05, that would exist between the two variables.

If more mind-numbing analysis is desired, then one can conduct more advanced statistical research using confidence intervals, hypothesis testing, chi-square analysis, Benford's Law (my favorite), large subsets test, relative size factor test, same-same-same test, correlations, and time-series analysis. All of these in-depth tests will create a deeper understanding of the data and overall fraud picture; however, as noted, these require additional time resources and also proper training in order to accurately interpret findings.

If we bring our attention back to the sample vulnerability worksheet we have completed, we can make several inferences. When we take a global look at the fraud vulnerabilities that exist within our respective companies on this sheet, we can employ basic data interpretation techniques to assist us with interpretation and application. We can immediately view all of the vulnerabilities on the left-hand side that we are challenged with; this will assist with current policy decisions or serve to help benchmark future decisions. We can see that all vulnerabilities have a high likelihood of occurring (10), telling us that we may have multiple risks that could compete for our attention. The significance of all risks is quite high (10, 10, 9, 5), with the exception of policy misrep—mileage at a 5; thus we may want to consider making this a lower-priority item. Our attention then moves to the last column, where we see that we have a significantly low value (1, 2) for both policy misrep—mileage and policy misrep—residency in our current fraud controls effectiveness. These extreme values would indicate to the reader that this may be the area that we can have the most impact in and one that would warrant our attention.

Again, it is important to remember that a highly focused strategy toward one specific fraud vulnerability is more effective and will have more impact than a slightly focused strategy toward all vulnerabilities. Companies that create highly focused efforts in one risk area have a high likelihood of seeing positive results, and my experience is that these impactful results in one area will ripple into other unforeseen fraud areas. As stated earlier, any system of fraud prevention is better than none!

Thus, the worksheet was developed to fit a niche of focused strategies with minimal time investment and to provide a robust, insightful, and in-depth overview of the state of fraud within an organization. The goal when completing the worksheet should be to identify multiple areas of vulnerability and then hone in on one or two specific vulnerabilities where an impact can be realized. Again, we cannot keep all the balls in the air, but we can focus on not dropping the most important ones! The most successful companies I have consulted with adopt this perspective; they target their resources on only a few of the problematic areas instead of a generalized approach to many areas. Once we have identified the area of focus for counter fraud strategies, then the fun begins. Developing red flags is the next step in translating the findings of the worksheet into a workable strategy.

Developing Red Flags

I have found very surprising details in regard to insurance companies and their use and development of red flags. It appears that almost all companies have a red flag document, but only about one-third of those companies frequently monitor, modify, and are in tune with this itemization. The other two-thirds are not familiar with the origin, and place limited effort or time into their use and application. One company asked me to assist with the development of a red flag protocol. When I asked for their existing document, they admitted that they use a red flag document from *the industry*. When I reviewed the document with the fraud director, it was obvious that many of the flags had no worth to the company. Many of the flags were for a totally unrelated line of business! Another company had an internally developed red flag document but informed me that its marketing department developed it; he confirmed that his fraud unit was not involved in the process at all.

A red flag document can be an actual script that is formally published. However, communicating red flags does not have to be a one-time document; I find great results when I work with companies to integrate them into a weekly, monthly, or semiannual red flag script, similar to an alert system or *what's trending* internal fraud publication. Circulating these red flags is a great opportunity to keep fraud fresh in the minds of the claims, underwriting, and legal departments. Something that is frequently overlooked in a counter fraud strategy: presence! A recommended tactic would be to rotate the duties of the red flag communication among a fraud unit; managers, investigators, and analysts could all take an active role in maintaining these flags and then circulating them monthly in an internal fraud publication. Again, this approach is highly recommended, as it serves two important purposes. It increases presence and continues to develop rapport and relations with other internal departments, and it also serves to communicate and circulate fraud trends and patterns.

The development of red flags is the second step toward developing highly effective counter fraud strategies. We can base our red flag strategy on the results from the vulnerability worksheet; almost all areas of the worksheet are ripe for the picking when it comes to developing these flags. We would start on the far top left of the worksheet and then methodically work our way down the sheet until we addressed all potential areas that could be included as a red flag. It is very important to mention that these flags are not direct indicators of fraud; these should only be used as a basis for further investigation. In our sample vulnerability worksheet, our red flags would start with the basic four vulnerabilities we identified and then they could be expanded or restricted as needed. Here are a few sample red flags that could be developed and used based on our vulnerability worksheet.

Policy misrep—residence:

- Policy zip code and policyholder area code are in different areas.
- Insured is employed more than __ miles (reasonable commute) from the policy address.

- Accident is more than __ miles from policy address.
- Insured is reluctant to meet appraiser/adjuster/investigator at the policy address and prefers a work location, coffee shop, and the like. This indicates that he/she does not actually reside there.
- Insured agrees to meet appraiser/adjuster/investigator at residence address, but needs more than a reasonable amount of time to get to the location. Again, this indicates that the policyholder does not actually reside at the address.
- Insured uses a PO box as his or her primary mailing address.
- Insured cannot produce documents that show his or her name and the policy address—documents such as utility bills, pay stubs, tax returns, or credit card bills.

Policy misrep—mileage:

- Insured is reluctant to release mileage information via phone or in person.
- Insured avoids appraiser/sales agent/investigator physically viewing the vehicle.
- CARFAX or other database check reveals a significant amount of miles in a short period of time. This would indicate potential commercial use.
- Accident occurs with multiple passengers that are not related or acquainted with the insured. This would indicate that the insured uses the vehicle for commercial or livery use, and potentially the mileage is misrepresented.

Other miscellaneous underwriting flags would be:

- Insured pays cash for premium.
- If insured pays via check, the checks are *starter* checks.
- Reemployment, the insured is very vague about his/her job duties and so forth.
- Premiums seem very high for the insured's socioeconomic status.

Both vulnerabilities of billing for services not rendered and up-coding will have almost identical red flags; thus they will be listed together.

- Insured's account of treatment does not match with the bills submitted.
- The bills are submitted in a boilerplate style; that is, the codes, dates of service, and other pertinent information are exactly the same for multiple patients.
- Lack of supporting documents, such as reports and narratives, submitted with the bills.
- Bills are submitted with very limited information on the patient.
- Medical facility cannot provide a patient sign-in sheet or other verification of the insured's presence.
- The medical provider has been flagged previously for questionable billing.
- Resubmitting a previously denied claim.
- Billing dates of service on a weekend or holiday.
- No diagnosis codes listed on the bills.
- Billing on consecutive days for a seemingly minor injury.
- Lack of supporting documents.

The above flags are samples for what could be integrated as part of a formal, or informal, counter fraud red flag strategy based on the vulnerabilities we identified. Below you will find a comprehensive list of potential red flags that could be used, in part or whole, in many other areas within an organization.

Red flags to be communicated to first notice of loss report takers:

- Insured provides a PO box.
- Multiple people making the loss report.
- Agent documentation indicates that the insured had a higher than normal number of inquiries to the agent prior to loss.

- Insured is hesitant to release personal information and is generally uncooperative.
- Insured is overly aggressive for a settlement and could also threaten contacting supervisors, state consumer bureaus, attorneys, and so forth.
- Insured seems unusually familiar with insurance terms.
- Multiple recent changes in policy coverage, including the coverage that specifically affects the claim being reported.

Red flags for disability/workers' comp/out of work claims:

- Claimant signature is similar to the doctor who provided disability.
- All disability verifications come from the claimant directly.
- Claimant can only be reached via cell phone during the day. This would indicate that the claimant is actually working while collecting.
- Consistent refusal to attend independent medical exams.
- Claimant avoids using the mail. This is to avoid federal charges if prosecuted.
- Multiple prior claims, potentially with the same doctor, attorney, and so on.
- Incident date is Friday and reported Monday. In workers' comp cases, this would indicate that the accident did not occur at work, but over the weekend while taking part in a personal activity.
- History of soft tissue injuries.
- No witnesses to the work incident (workers' comp).
- Inconsistent statements to incident (workers' comp).
- Significant change in the employee's work status (disgruntled, facing layoff, soon to resign/retire, seasonal employees).
- Questionable documents submitted (forged out-of-work document, forged prescriptions).

- Activity inconsistent with injury as a result of activity check or surveillance.
- Accident is not reported promptly.
- Poor attendance record prior to incident.
- Disciplinary action prior to injury.
- Problems with coworkers or supervisor.
- Incident occurs in an area that the employee does not normally work.
- Employee missed or passed over for promotion.
- New employee.
- Injured worker has a history of short-term employment.
- Injured worker mentions to coworkers a prior medical condition.

Red flags for sales staff/sales agent fraud:

- Significant changes in a specific sales agent or sales office in the following areas:
  a. Agency sales increasing or decreasing.
  b. Customer complaints.
  c. Number of new policies written.
- Policy address the same as the agency address (this flag was the first indicator in the largest broker investigation of my career).
- Frequent lapses in coverage for the policies the agent manages.
- Agent has high frequency of SIU involvement of policies written.
- Agent known to deal in cash; cash visibly seen in office.

Red flags for life insurance/death claims:

- Life insurance application has many gaps in information—lack of detailed health history, doctor information, and the like.
- Application is not signed.
- Insured writes an earlier date of birth on the application.
- If a death claim, it occurred out of the country.

- Autopsy report shows a different height and weight than on record for the insured.
- Autopsy shows inconsistent dental records.
- Insurance amount requested at inception appears extremely high considering the insured's demographic.
- Death reported soon after policy was written.
- Premium paid in cash.
- Beneficiary is in dire financial condition.
- Beneficiaries are pushing for a fast settlement.
- Cause of death ruled undetermined, accidental, or self-inflicted.
- Remains not able to be produced.
- If international, body disposal violates local traditions.

Red flags for personal property submissions:

- Claim submitted for a large amount of cash.
- If a theft, there are no signs of breaking and entering.
- Claim submitted for multiple important and valuable family heirlooms.
- All claims submitted are for items that are the most expensive model/brand available.
- Lack of receipts or formal documentation for items claimed.
- Receipts and/or documents that look suspicious (no dates, no monetary amount, etc.) or the writing is similar on multiple receipts (insured wrote the receipts).
- Receipts are photocopied, yet the insured cannot produce the originals.
- Insured willing to negotiate the value of the items.
- Items from the insured recently put up for sale.
- Recent changes in policy coverage that benefit the insured with the value of the claim.
- Insured submits claims for items that seem above their means.
- If art or jewelry, does the appraisal look suspicious?

Red flags for claims on commercial properties:

- Did event/loss occur when the security cameras or alarms *mysteriously* were not working?
- Insured's business is financially in jeopardy: declining sales, terminating employees, and so on.
- No witnesses to loss.
- Policyholder background information that would indicate he/she was looking to get out of the business: retirement, illness, etc.
- Loss occurred on a weekend or a holiday.
- Loss occurs at night.
- Claims are submitted for seasonal equipment, such as lawn mowers at the end of summer, snowshoes at the end of winter. This is to avoid costly inventory.
- Inventory old and in need of updating.
- Deteriorating area surrounding the building.
- New competition close to the insured's business.
- Items removed from property prior to loss.

Red flags—arson for profit:

- Fire occurred on a holiday, weekend, or late at night.
- No witnesses.
- Insured's pets not present at the time of fire.
- Insured's important papers missing.
- Combustibles present in building or in storage close by.
- Building or business recently purchased.
- Multiple fires occurred.
- Origin in an unlikely location—bed, for example.
- Insured in dire financial condition.
- Insured's activity is unconfirmed at the exact time of event.

Red flag development is the critical second step in the counter fraud process, a step that organizations should take seriously in order to ensure highly effective strategies. The red flags listed above may be used and applied directly in many different fraud scenarios, or they may be used as a foundation for further fine-tuning. Now that we have a deeper understanding of red flag development, we will transition into the third step in the counter fraud process: integrating controls. In step one, during the vulnerability assessment, we came to understand what areas within our organizations are the most vulnerable. In step two, the red flag development, we translated those vulnerabilities into more focused categories. Now, in step three, integrating controls, we will learn how to start to apply those red flags into workable strategies for preventative efforts, strategies that have been proven effective in many of the consultations I have been involved with during my career.

Integrating Controls

At this phase in our vulnerability assessment, we have a clear idea of what areas need attention, but we may still be uncertain as to the method to reach the most beneficial result. In many organizations, time and resources are spent to reach this point, but many fall short, as there are unfocused strategies for development. If strategies are not integrated, then this will be seen as a vulnerability and, as we learned from our discussion on criminology, will result in more fraud occurrences, as the criminal element will be drawn to these gaps in strategies.

As we have stated several times, one of the main messages of this book is to stress the importance that organizations install any sort of system of prevention, no matter how small or seemingly incidental. One of the most cost-effective preventative tactics is to leverage technology and take advantage of the many different advancements currently available. It is important again to stress that companies should work with whatever resources are available; I have witnessed effective anti-fraud programs utilizing internal IT staff and internal IT capabilities and also companies that utilize software fraud solutions and predictive analytics. Current fraud detection tools available on the market are extremely effective, as they tap into the incredible power of interpreting data. However, as stated, any system is better than none, so if internal resources are the only option, then workable strategies are possible. In the next chapter, "The Behavioral Bridge," additional concepts will be presented that could also be included in this Integrating Controls section; however, for organization purposes, these concepts are presented in the subsequent chapter.

While consulting on an international basis, I am directly involved in the initial consult, rollout, and modification of counter fraud software solutions of Risk Shield, a product by Inform. As I see such positive results with our implementations, I am, of course, prone to recommend intelligent software solutions as an effective counter fraud strategy, as these systems provide an incredible return on investment. Let us briefly explore the structure of these systems. The first step with these rule-based fraud detection systems is to conduct an in-depth, on-site consultation to identify key areas of focus, as we did with the vulnerability assessment. These key areas will then be translated into rules that will be the basis for the software detection system. These rules will be used to score all incoming claims and assist with easy fraud identification. The goal of this program is to collect, connect, and analyze data using social network analysis, fuzzy logic, dynamic profiling, pattern recognition, and predictive technologies. These systems use models, logic, and algorithms to quickly adapt to new modus operandi of fraudsters. Here are some basic definitions that are common in the software detection world.

Predictive Analytics

- The analysis of historical data to try to make predictions about future or yet unknown events.
- Understanding claims history is part of the predictive scoring.
- Dynamic profiling takes into account what has happened in the past regarding the policyholder, the third party, or any other involved entity.

Fuzzy Logic

- Variables may be any real number between 0 and 1 (i.e., not suspicious, somewhat suspicious, suspicious, very suspicious).
- Thinks like a human and takes exonerating facts into consideration.
- Like a human, system uses vague (fuzzy) terms such as *recent policy issued, expensive car,* and *high claim amount.*
- Combined with dynamic profiling, system is self-learned to predict trends and unseen patterns.

Social Network Detection

- Finding relations between entities of different claims, such as bank accounts, credit cards, addresses, phone numbers, IP addresses, and other tracked data.
- High value in finding organized fraud groups.

Text Mining

- Analyzing free text to gather additional information to include in analysis.

Data Mining

- Searching data to find new patterns or specific information.

External Data

- Additional third-party information about customer, claims, payment methods, and many others.

Let's delve into a specific internal control example. One common scheme in medical billing fraud is duplicate charges, which occurs when an unscrupulous medical provider intentionally submits two bills for the same date of service in a very creative manner. If a company does not have an internal control (data detection system), then this provider could be paid twice for the same bill. For example, a provider would submit a bill on August 19 that covers two weeks of treatment; from August 1 through August 12. A second bill is sent on August 26, which also covers two weeks of treatment, from August 8 through August 19. Notice that there are duplicate charges in the entire week of August 8 through 12. If an organization does not have an internal control to identify and filter this duplicate week, then this will be paid!

Other billing scams can come in the form of billing for services not rendered. It is vital to make sure the diagnosis codes on the medical bill are consistent with the medical procedures actually being performed. I have witnessed many fraudulent bills that were submitted where the procedure and actual injury were separate areas of the body, such as when a patient with a forearm injury had several bills submitted for expensive stomach treatments; again, if no internal controls are present, then this will be paid. Questionable providers will also unbundle bills, which occurs when they bill procedures separately for higher-billed amounts when the procedures should all be grouped together for a lower-billed amount. These can easily slip through to payment if no controls are present.

There are other forms of technologically based internal controls, such as social network visualization or link analysis. These programs are highly effective at discovering fraud rings using interactive visualizations. These networks may be hidden if a company is only looking at a single claim. These hidden patterns will help to identify staged losses, relationships, and show a history of prior claims. Below is an example of a social visualization program.

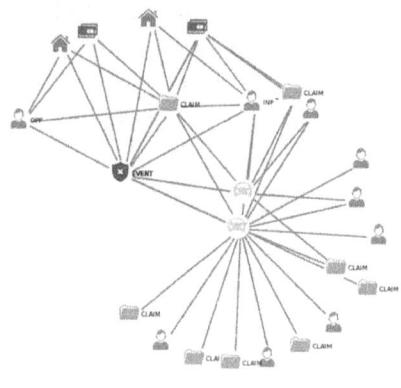

A link analysis can show patterns and relationships between the following:

- social networks
- addresses
- vehicle identification numbers
- e-mail addresses
- IP addresses
- fax numbers
- cell and landline phone numbers

Thus far we have explored integrating controls using outsourced software detection systems. But what options are available with limited resources and a minimal, or no, budget? This will be the focus of this section of the analysis. If there are no options for outsourced solutions, companies can utilize internal resources, and creativity, to build a system that works for their unique challenges.

There is no doubt that use of data is the trending wave of the future of fraud fighting, with big data and the Internet of Things a constant theme at conferences, in publications, and in all forms of formal and informal conversations. Companies with limited resources must focus on methods to use data as a strong fraud-fighting tool, as it will provide the most return on their time investment and provide valuable information for application. Organizations that see the most success with minimal resources first conduct an internal assessment on what data is available. It is also recommended that fraud units canvass internally, in all departments, and determine which employees have hidden talents and credentials that can assist with this endeavor. One company I worked with found an ex-IBM consultant working in its underwriting unit; another company found a property damage assessor that had a hidden talent for building websites. Both employees were instrumental in the development of these internal systems. I also find that many employees from outside the fraud units are extremely willing to assist, as they find intrinsic value in fighting fraud.

Once internal resources are identified, the next step is to determine what data is available for use. This is where collaboration with internal IT is very important, as many companies have different systems for different areas of the insurance operation. One company I worked with had an internal system that could extract information from the first notice of loss from the sales staff; this was then imported into simple Excel and Access programs for interpretation. An analyst would then delve into these programs looking for many of the underwriting flags mentioned earlier, such as claim on new policy, unlisted driver, and so on. If the IT skills of internal staff are more advanced, then I have seen where these flags can be a part of an underwriting risk score, similar to a rule-based software system.

One of the most significant challenges in the use of internal data is that they are often housed in different areas, in locations that do not talk to each other. I recall one company that had six different systems; in this scenario, the internal solution is a bit more challenging but not impossible. This carrier focused on the most important system, the one that processed its medical bills, as this is where it had the most vulnerability. In this case, intelligence gained from conferences, networking, and other fraud units was used to identify key unscrupulous medical providers, and then the TIN numbers of these providers were used as a filter in Excel and Access. The analyst would then be able to locate the claims associated with this TIN and issue an alert to the claims department in a manual form, similar to a weekly newsletter. However, internal IT implemented an even more advanced step by accessing the billing system and sending an alert to the claim representative when a bill was received with that specific TIN, thus prompting a fraud referral.

We have explored integrating controls using data, but what about human resources? I have found that companies are extremely reluctant to outsource any function, as it is seen as an opening for future replacement of staff, including the internal individual approving the outsourcing. However, my experience has been quite the contrary, as companies that consider outsourcing as a supplement, not a replacement, for current staff have seen very impressive results. Using outsourced companies and experts on a selective basis in specific areas can augment and streamline many areas of the fraud process. These outsourced experts can offer assistance in the areas of hiring, consulting, training, investigations, IT, workflow optimization, analysis, management, medical, biomechanics, appraisals, and many other areas. Companies operating on a limited budget can see immense returns by using these experts to add value to existing systems. A specific company comes to mind that utilized one external investigator to handle all of its complex medical investigations. The return on investment for this single outsourced individual was incredible, as it allowed the company to investigate very damaging medical cases, cases that it did not have the internal staff to handle previously.

The academic research shows extremely strong support for the utilization of technology in all forms of counter crime prevention, including fraud. Criminals have realized that leveraging technology is the way of future criminality, with alarming statistics that show incredible growth in crimes utilizing these advancements. We, as fraud fighters, should also embrace this trend and use these great resources to our advantage. We now have an understanding of the benefits of utilizing technology. However, there is an often overlooked aspect of integrating technology, and that is the human component. My research and professional observations have revealed that some companies that utilize technology may have a false sense of security that once the technology is installed, that technology is all that is needed in order for their organizations to be *safe*. That is, they lack the full utilization of the human component.

The secondary research question explored in my dissertation study was: What preventative tactics have been utilized by insurance fraud investigators, and have they been effective? The sample population felt that technology was one of four major tactics that are the most effective at fighting fraud. One of the participants felt that "most of the effective systems are the use of red flags or indicator systems to provide notification of a potentially questionable claim." Another participant stated that one needs to be technologically savvy to fight fraud and that big companies "always have a good infrastructure to fight this stuff," but noted that smaller companies are at a strong disadvantage, as they don't have the resources to support this. One of the consistent themes among the sample was the benefit of predictive analytics and software detection systems. As stated by one participant, "This is big, one of the most effective tools we have used in a while."

As stated, the application of technology is a critical factor in fraud fighting. However, almost all members of the sample population warned that the sole use of technology, without a human component, is absolutely ineffective. One sample member provides an accurate summation: "The information coming out of the system is nowhere sufficient to deny a claim or conduct a prosecution, it is more a set of indicators saying that a human needs to get involved here, to look things over and then make a determination."

A large majority of the sample discussed a two-pronged approach, whereby an indicator system filters through all incoming claims (first prong), and then these files are reviewed (second prong) by a fraud analyst or claims liaison. One member stated, "No matter how far along we are, humans have different capabilities than a computer, and at some point this information needs to go through a human filter."

The *human component*, as we have so termed it, is an often overlooked area of focus for many organizations. I have seen countless companies install technologically based internal control systems and then adopt the mind-set that all the work is complete, thinking that the technology alone will handle their fraud issues. This is a critical error, and is a very simple one to rectify. All systems need someone to be able to decipher and interpret the data that is being generated from the internal control system, to address the outliers, false positives, trending areas, highs and lows, and to be able to take a global view of the results for application. A common approach toward the integration of the human component into controls is the development and use of analysts or liaisons. These individuals have the full- or part-time responsibility of reviewing data, using both macro and micro lenses, and making inferences and modifications based on the results. If the quantity of data does not warrant a full-time analyst, than I have seen companies operate very successfully using a part-time model. If an investigator or claim representative has a talent for technology, then this could be a great opportunity to allocate this individual to work one or two days a week in the analyst or liaison role.

In my dissertation sample, a large percentage of the survey participants had great results with use of a fraud analyst, an employee who is office-based and who reviews databases, red flags, computer programs, and files looking for patterns, trends, and outliers. Almost all of the participants thought that utilizing an analyst was the most effective strategy to deal with decreasing fraud budgets and other challenges. One sample member stated, "We see a huge return on investment with our analyst." Another participant remarked, "Remember—we want to be proactive, not reactive. If it's a closed claim, it is almost impossible to get money back. This is the wrong approach. You need to stop the money before it gets paid, on open claims, review them; our analyst is poring over hundreds of claims per month and pulling the ones they feel should be investigated based on our guidelines, our indicators; we are hitting the ball out of the park with this approach."

Inserting the human element into fraud detection can be done on many different levels. As we have discussed, a full- or part-time analyst embedded within the designated fraud unit is a very effective strategy. Including the claims department in fraud detection has also shown to provide incredibly positive results; as these claims units are the first line of defense, it benefits fraud units to develop strong working relationships with them. A claim liaison position is one that carries the responsibility of filling a vital gap that exists between claims and fraud units. The liaison will serve to be the point of contact for all claim representatives that have questions, issues, or are in need of assistance on fraud-related matters. He or she will also communicate information back and forth between the fraud units and claims departments and act as a more familiar *face* of SIU. In many companies, the claim liaison may be responsible for the review and transfer of cases into fraud units; in other companies the liaison may monitor and maintain a certain number of cases that are in SIU for investigation. This all depends on if the company maintains a protocol whereby the claims department still retains the master file, or if the entire fraud file is transferred to the fraud unit.

Now that we have an understanding that a human element to internal controls is vital for success, let us integrate some of the concepts we learned on criminological theories from chapter 2 into our analysis. One of the criminological theories we discussed at several points in this book, and one that is applicable in a fraud setting, is routine activities theory. To review, routine activities theory states that for crime (fraud) to occur, three critical elements need to be in alignment: motivated offender, lack of capable guardian, and a suitable target. If we apply this theory in a fraud scenario, we assume that we have a motivated fraudster and a suitable target (insurance company); then according to routine activities, if a lack of capable guardian is present, then fraud will ensue. Thus, if we control the presence of the capable guardian, then fraud will be diverted. In other studies, as discussed earlier, where routine activities has been tested, a capable guardian can be anyone that is seen as a figure or has a presence in the fraud setting. In our fraud scenario, a capable guardian could be an insurance agent, claims adjuster, fraud investigator, analyst, police officer, attorney, or anyone that is seen by the fraudster as a gatekeeper.

One dissertation participant confirms this contention and states, "One of the most important strategies to fraud prevention would be more focused claims and fraud staff, staff that would specifically focus on fraud prevention as its core accountability."

Integrating controls is a critical step toward developing highly effective and capable counter fraud strategies. These controls can utilize technology; however, studies have shown that a human element must be inserted in order to glean the most benefit from our efforts. Creating full- or part-time analysts or liaisons has proven to be an effective strategy for many organizations, as this fills a communication and rapport gap that often exists between claims and SIU. Furthermore, the absence of capable guardians is a core component of routine activities theory, a theory that has proven itself in many criminal scenarios. We have demonstrated that capable guardians, or gatekeepers, should be inserted into this counter fraud formula, as they will show a presence to criminals and serve to deter potential fraud.

As we continue to explore applications of integrating controls with a limited budget, it is common to look for new, trendy, and contemporary strategies that utilize the latest techniques. However, in my opinion, one of the most important tactics that can be applied is very simple and a back-to-basics approach, and that is one of communication. It is very surprising how increasing communication between units and departments can have significant impact, with little or no additional resource allocation. One very large global carrier comes to mind that utilizes basic communication as an integral part of its anti-fraud strategy. This carrier is very capable of purchasing advanced software and utilizing the latest tech tools; however, it relies on strong communication between units as the core of its fraud strategy. In the workers' compensation line of business, it is understood, through solid communication between claims and the fraud units, that the most powerful indicator of suspicious activity is when the claims unit, or first report unit, cannot reach the injured party at his or her home number during business hours. This implies that the injured party is in fact working another job while collecting workers' comp benefits. This is considered the strongest red flag of potential fraudulent activity in this line of business, according to the analysis. Accordingly, the claim representatives and first report takers will notify the fraud unit if they attempt two calls with no success to the injured party during daytime hours. This then generates an automatic referral. This entire process is manually based, but all based on strong communication between different departments, created by a claims/SIU liaison and weekly collaboration meetings.

A final step in the vulnerability assessment process is to monitor and modify—that is, to analyze the data and information we have available to us and then modify our systems accordingly.

Monitoring and Modifying Controls

Internal controls are a necessary component of an effective counter fraud scenario, but we also must have a protocol for monitoring, reviewing, analyzing, and modifying these controls based on our review of the data. We must be able to review and then comprehend the information in order to look for trends, patterns, and outliers in order to modify the controls toward a more favorable result. Monitoring our controls can be as complex or as simple as needed, and can be calibrated with the human and financial resources that we have available to us. It is more important to have a simple system of monitoring and modification than no system at all!

Many insurance companies and agencies that utilize fraud solution or case management providers have the ability to leverage technology to assist with the monitoring and modification process. These monitoring programs—also called dashboards, business intelligence programs, and management reporting systems—work in conjunction with other internal data systems and have the ability to filter information based on user need. Below is an INFORM program, named Business Intelligence, that shows a visual view of data pulled from a carrier's database; this is extremely valuable so we can start to develop an understanding of business and fraud trends and patterns and see which areas we need to focus on and which ones we can withdraw resources from. These technical systems offer the user the ability to modify what is seen in the final visualization, whether it is claim volume, fraud volume, geographic focus, exposure, threshold, referral percentage, line of business, and an almost infinite number of data sources.

This visual representation, and the ability to alter data fields, is an extremely effective method as we start broad and can then pinpoint areas of opportunity. We can also easily view outliers, or data that falls outside of the norm, for deeper analysis and interpretation. In one consultation, we decided, after initial review of multiple data fields, that underwriting fraud appeared to be more prevalent in a particular geographical area. We then decided to further explore this and cross-referenced those fraudulent cases with the sales agent that wrote the policy. Surprisingly, we immediately could see a very significant relationship between these questionable policies and one specific sales agent. The fraud unit opened an investigation into this agent, which ultimately resulted in a very large-scale case that led to prosecution.

PSYCHOLOGY OF FRAUD

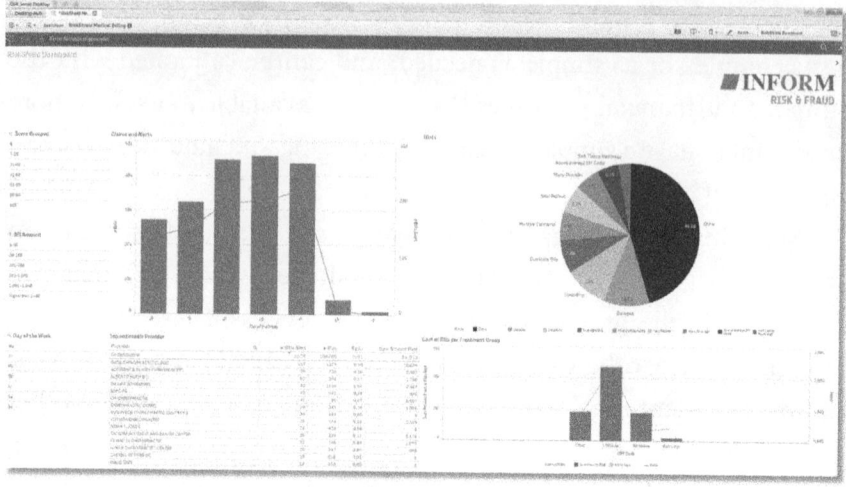

Many fraud detection programs can utilize rule-based systems to assist with filtering and scoring claims. It is vital that carriers that have these systems in place monitor the results of the scoring and adjust the rules as necessary. Reducing the number of false positives is the primary goal with these systems in order to fine-tune fraud detection capabilities. Below is a slide where we see a visual depiction of rule tuning, which is the specific process of monitoring and modifying rules in order to focus fraud detection efforts. For example, in one carrier, a rule was installed that scored claims on policies that were incepted within thirty days. Upon review, it was realized that because that carrier's particular book of business was subpar, and therefore prone to a higher number of claims on new policies, this rule had to be adjusted to claims that occur within ten days of policy inception. This resulted in more focused scoring and a higher quality of claims being reviewed by the fraud unit.

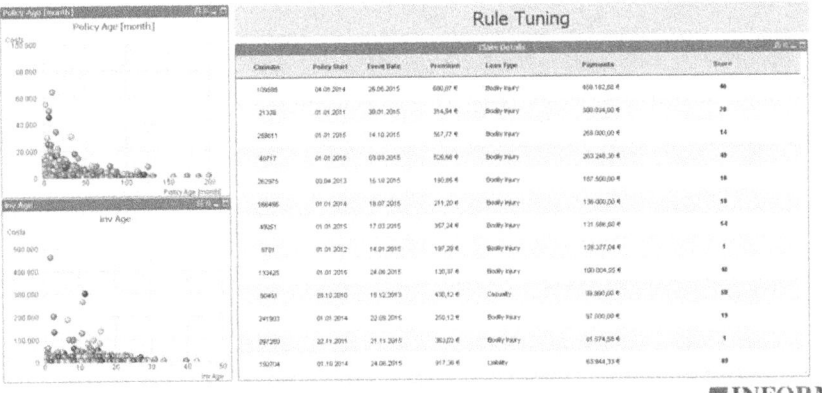

RiskShield Business Intelligence

■INFORM
RISK & FRAUD

Thus far we have explored methods to monitor and modify using advanced technology. How about those carriers that do not have the resources to leverage such high-level technology? Again, the message to carriers is that any monitoring and modification is better than none. I have worked with carriers that use seemingly basic systems, such as Excel, Access, or other corporate-specific internal systems with success. The key to effectiveness with these carriers seems to rely on the ability to utilize the internal talents of existing employees. It always seems to surprise carriers how many employees they have that have hidden talents. In one carrier I worked with, a claims representative had a college degree in programming and was a computer consultant prior to joining the insurance company. He was a key element in the development of that company's counter fraud program, and they used only internal resources. He was able to talk the computer language and translate the wishes of the fraud unit into a workable fraud detection system. As stated, work with whatever resources are available to you, even if that means relying strictly on internal personnel and systems.

The modification process also should include integrating intelligence gained from outside sources or networks. This is another area of fraud detection that is often overlooked by carriers: networking. The secondary research question posed in my dissertation study was: What preventative tactics have been utilized by insurance fraud investigators, and have they been effective? Four themes emerged from this question: (a) fix the social problem, (b) more focused and specialized staff, (c) technology, and (d) networking. Let's explore networking in more detail.

Networking was considered among the dissertation sample as one of most significant tactics to fraud prevention. In an environment of focused budgets, this was seen as an extremely cost-effective strategy to increase the foundational knowledge of fraud strategies and help to focus efforts. One sample member replied, "Learning what others are doing helps to see the big picture, where to focus efforts and where to let things go." Another sample member, who was employed by a smaller carrier, stated that almost all of the carrier's referrals into SIU are based on intelligence and knowledge gained from networking with other investigative units. The sample also agreed that one of the most effective ways to network was to attend seminars, meetings, and conferences by professional fraud-fighting groups. One participant discussed how the registration fees for these meetings were well worth the investment: "You get that back one hundred times over with the intelligence shared, the networking, and the contact lists from these meetings." The participants agreed that attending these meetings was the most cost-effective tactic to network and increase fraud intelligence for application.

Monitoring and modifying internal controls is a critical step in developing strong counter fraud programs, a step that can be accomplished by using advanced, outsourced technological solutions or existing internal capacities. The monitoring and modifying process is necessary in order to adapt to the constantly changing world of fraud prevention, a world that requires carriers to be fluid and flexible in their approaches. Another critical step in developing highly effective counter fraud strategies is for organizations to reflect on what measurement parameters are specifically used in their current fraud programs. This topic area will be explored in depth in the next section.

Analyzing Measurement Criteria

We have discussed how the measurement of fraud is one of the barriers to accurate fraud efforts. As discussed early in chapter 4, my dissertation sample reported that one of the primary barriers to current efforts in fighting fraud is that fraud is difficult to quantify. The sample reported overwhelmingly that the measurement of fraud is very disjoined and that because there are no consistent approaches toward measurement, there is a lack of awareness of the true fraud problem.

Earlier in the book, we discussed the rigor required to publish research at the peer-reviewed level; furthermore, mention was made of how critical our academic peers are when we attempt to publish a research article. Coming from an investigations background, I understand the parameters of being *critical*. However, it wasn't until I entered academics and received formal training from highly accomplished researchers that I truly came to understand how to take this critical approach to a much deeper level. This critical eye should be part of all analyses of counter fraud efforts, including how we are conducting our measurement and collecting our data.

In order to successfully monitor and modify a system of controls, one must have the correct measurements in place. Quite simply, we must be measuring the right thing! As we have discussed, flexibility with counter fraud efforts is needed in order to adapt to changing fraud schemes, and as such, the measurement parameters also need to be modified routinely as well. If our measurements are focused on arson rings in the automobile line of business, but current trends are showing spikes in commercial/home arson, then we must adjust our parameters accordingly. In addition, we must understand that in order to be accurate, our measurements must be consistent. Retrieving and reviewing data from only a few months prior will not provide the information needed to accurately assess our system of controls. We need to be sensitive to this and understand that oftentimes we need to participate in a cycle of data collection (which could be a year, or more) in order to gain meaningful data for interpretation.

Inconsistency in data collection is also an important area to consider when discussing fraud trends with peers. I recall a professional board meeting I participated in during my career. The topic of discussion was focused on trending fraud issues in a particular state in the Northeast. When I reviewed the data that was reported from each organization, I immediately realized a significant problem: inconsistent terms of measurement. I randomly asked several agencies how they determined what defined insurance fraud in their area; surprisingly, none of the agencies had the same measurement definition. One defined fraud as only when successfully prosecuted, another when certain thresholds are met, and so on. Due to this lack of calibration of terms, the data was virtually useless from an application perspective.

The main message to be delivered in regard to measurement is to take a critical view of how data is collected and what specific parameters of measurement are used. Information can be skewed, and the resulting data will provide a different view of the direction for effective counter fraud efforts. Data is the source of all effective preventative efforts, but be guarded in how it is collected before application.

In this chapter, we began to apply the criminological theories that were presented earlier. We explored the steps of developing an extremely strong counter fraud effort, which involves identifying risks and vulnerabilities, assessing vulnerabilities, developing red flags, integrating controls, monitoring and modifying controls, and, finally, analyzing measurement criteria. The vulnerability assessment was reviewed and is a procedure that companies and departments from all sizes, lines of business, and locations can utilize as part of their fraud strategy. In the next chapter, we will explore how companies can develop a multipronged approach to prevention, an approach that also integrates the criminological theories discussed in chapters 2 and 3.

CHAPTER 5

Developing a Multipronged Approach to Counter Fraud

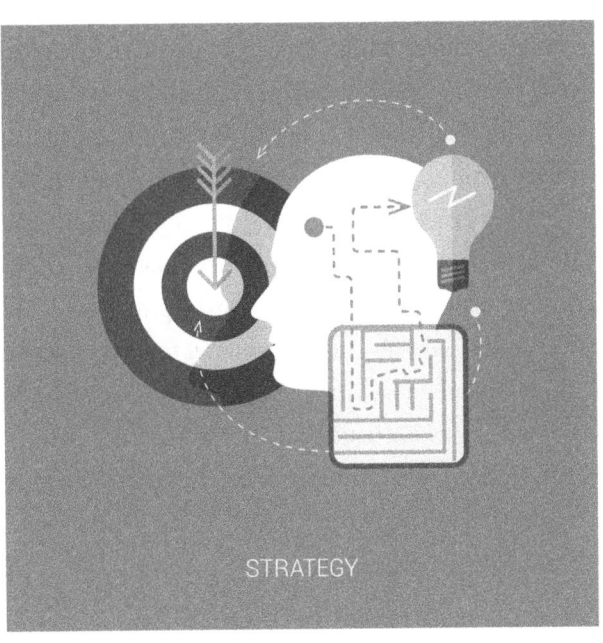

In this chapter of the book, we are going to merge all of the topics, concepts, and ideas discussed thus far and look to develop a deeper understanding of how psychology and counter fraud efforts intertwine. A summation of criminological theory in chapter 2 provided us with the perspective needed to understand the mind-set of the criminal and what motivates deviant behavior. We then delved into various specific strategies to prevent fraud, all focused on ways to reduce the opportunity to commit fraud, the underlying goal. The main focus of this final section is to illustrate how organizations must develop a multipronged approach to fraud prevention, as this is the only method to effectively cover all possible areas of vulnerability within a company or agency. We will discuss why the mind of a fraudster will be affected by different counter fraud efforts, thus requiring agencies to take multiple approaches to fraud prevention to be truly effective and have a deep impact.

Psychological Profile of a White-Collar Criminal

We know from existing research that white-collar criminals are very unique as compared to other types of criminals. These offenders are characterized as having higher levels of intelligence than other street criminals and are often grouped with computer hackers, burglars, and terrorists on cognitive measurements. White-collar criminals are known to be unpredictable and stealthy in their approach to criminality, thus making them very worthy of extremely strong and focused strategies for prevention. From a demographic perspective, they are most often male, have a low degree of self-control, high degree of hedonism, and above average narcissistic attitude (Brody and Perri 2016; Ragatz and Fremouw 2010). When we use the term white-collar criminal, we most often think of a corporate-level crime involving CEOs and upper-level management involved in insider trading, blackmail, and counterfeiting.

Fraudsters have been grouped into the category of a white-collar criminal as they appear to carry many of the characteristics of this type of unique criminal, and second, there is a lack of information on the psychology of the insurance fraudster. Thus, we can make certain parallels and connections with the research performed on psychology of the white-collar criminal to that of an insurance fraudster. I have seen several intriguing studies that have found an interesting relationship between insurance fraud and tax evasion from a psychological perspective, showing similarities in the personality traits of these offenders. These commonalities will be discussed in a later section.

When we approach discussions on specific counter fraud programs, we need to keep in mind the main theme of this book: the psychology of a fraudster. In earlier discussions, we presented the two types of criminals that we are faced with: the opportunistic and organized criminal. However, the research, and my own personal experiences, have revealed that there are many different personality types within those two broad categories, personalities that will be affected differently by counter fraud approaches. The psychology of each fraudster may be different; thus, as we develop our agency-specific fraud programs, we need to keep this in mind and develop this multipronged strategy.

Let us consider the psychology of a typical fraudster that would be grouped as an organized fraudster. This individual considers fraud a business opportunity and uses well-contrived schemes and scams to create income. In dealing with these larger-scale criminals, one would assume that the most effective counter fraud approach would be data analytics, predictive modeling, and patterning; basically, using technology to flush out these schemes. However, I recall an experience with a dangerous offender that proves otherwise. One of the benefits of being an academic researcher is the opportunities that are oftentimes presented. Several years ago, I conducted an in-depth interview with a convicted serial arsonist, a criminal that defrauded millions of dollars, was involved in several attempted murders, and (great result) was prosecuted and charged federally. After he completed his prison sentence, he was willing to speak about his life as a fraudster, which provided a rare opportunity to get a glimpse inside the mind of a fraud offender.

John Doe's fraud schemes were first discovered by a federal law enforcement officer who used patterning and visualization programs to connect the main perpetrators in this large-scale organized arson ring. This officer spent almost a full year investigating and compiling a solid case for prosecution. There came a point when the officer gathered enough evidence and was ready to confront the subject. The officer made a phone call to the offender and was prepared for an in-depth, detailed discussion, containing aggression, denial, and uncooperativeness. To his surprise, John Doe immediately turned himself in to the agent and actually thanked him for intervening. So, yes, technology did play a factor in the final detection in this case. However, after I had the opportunity to interview the subject and was more familiar with the specifics of the case, it is my belief that he would have ceased his fraudulent activity well before his discovery had certain other countermeasures been in place.

John Doe was a hardworking father of three who resided in an inner-city, impoverished metropolitan area. He had a high school education and held various jobs, with no particular industry or trade focus. He worked in retail, construction, health care, and many other industries in lower-paying positions. He suffered a legitimate fire to his rental property and made a claim on his renter's policy for several personal items damaged in the incident. He made the initial report to his sales agent and then communicated very briefly with his assigned claims representative. He was provided a full review of the claims submission procedure and followed this protocol when submitting claims for approximately five items that were damaged in the loss. He had receipts for four of the items and sent in these documents with a claim form requesting payment for five items. Approximately two weeks later, he received a claim settlement check for all five items. When he reviewed the settlement check, he noticed that the check was for approximately $1,000 more than what he had claimed. Upon further investigation, he realized that he had mistakenly submitted a claim for the fifth item (without a receipt) for $1,200 and not $200. He stated that he pondered notifying the claim representative but was in very dire financial condition and decided to keep the money.

John Doe remarked at this point that he did face a moral dilemma and was very disturbed by his acceptance of the claim check with the additional settlement. He described himself as a very religious person and said he struggled with the decision, but in the end decided to retain the check and cash the full amount. As he went to work the next day, he stated he thought of how he *earned* more on the extra $1,000 in the claim check than he makes in a month. And this began a dangerous cycle of crime that resulted in dozens of fraudulent arson cases with multiple accomplices and multiple carriers—and over a four-year span totaling millions of dollars in exposure.

John Doe has a full understanding of how he was caught; he comprehends that numerical strategies and patterning are ultimately what led to his demise. The most insightful discussion with him came when we dialogued about other counter fraud efforts and which approaches may have deterred him from fraudulent behavior. We discussed many of my dissertation themes, which have been mentioned at various points in this book, such as honesty triggers, insurance companies are their own worst enemy, and fraud is difficult to quantify. I asked him to choose two or three that may have had an impact on his deviant behavior; he chose (1) honesty triggers, (2) public awareness, and (3) insurance companies are their own worst enemy.

John considers himself a religious (and honest) person, and he stated if there was an honesty declaration, or some honesty statement in any of the forms or documents he received, this may have made him internally reflect and may have triggered some moral compass. Second, he stated he was not aware of the damaging effects of fraud, both financially and from a violence perspective (in his staged arsons, he was involved in several violent crimes). He explained that a public outreach campaign presenting the dangerous effects of this crime also may have thwarted his efforts. Third, he explained that insurance companies portray large profits, and accordingly the public perceives them to have a lot of extra money at their disposal. He also remarked that the insurance companies themselves do not seem to be well informed, as he spoke to many internal employees in his targeted organizations, varying from sales to claims employees, that did not seem familiar at all with the fraud problem. He mentioned that much more internal training was needed to help see the true impact of the crime.

This inside insight into the mind of a fraudster is very enlightening and reveals to us many important perspectives. It serves to confirm one of the main tenets of this book: to utilize a multipronged approach at fraud prevention. Even though John Doe would be formally categorized as an organized criminal, due to the contrived, organized method of his fraudulent activity, his actions would potentially have been thwarted if other countermeasures were in place with the targeted carriers.

We all have different personalities and psychological profiles; each of us is motivated in different ways by different factors. In many of the collegiate classes I teach, I will often conduct a personality test on the students to further exemplify this point. I will administer the Myers-Briggs test, the Sixteen Personalities test, the Caliper Profile, Gallup Strengths Finder, or popular Minnesota Multiphasic Personality Inventory to assess each student's personality type. Prior research tells me that criminal justice students should naturally score higher in the areas of authority, ethics, leadership, and decision making under stress. However, my personal experience has revealed that there is no identifiable pattern in their results; they all have very different personality traits that vary greatly. If the research reveals that we all (criminals as well) have different personality types and psychological profiles, motivators, stressors, and so forth, then how would one specific counter fraud strategy deter us from criminality? It wouldn't. Thus the need for multiple approaches that will reach the highest number of potential fraudsters.

We can reference the Coalition Against Insurance Fraud's (n.d.) clustering of insurance customers to further exemplify this point. The coalition grouped customers into four main categories: moralists, realists, conformists, and critics. Moralists are ethical customers who strongly disagree with fraud and would never commit a fraudulent act. Realists have a very low tolerance for fraudulent activity but will justify it in certain circumstances. Conformists are generally accepting of fraud, as it is a common occurrence in our culture. Critics are those that have a high degree of tolerance for fraud and precipitate its occurrence. Accordingly, each of these four fraud types would be affected differently by different counter fraud efforts.

Realists are those individuals that we would term rational thinkers; that is, they understand the damaging effects of fraud, but also feel it is justified in certain scenarios. If we draw upon earlier conversations of the classical approaches to criminology, the rational criminal will weigh the benefits and risks of engaging in a criminal action. Therefore, preventative efforts toward these rational thinkers need to focus on the severity of punishment and the high probability of detection. Organizations should showcase zero tolerance and make examples of fraudsters caught and prosecuted in order to impact these psychological typologies. In the global view of fraud prevention, the strategies for realists may be considered softer than other approaches.

Conformists would be considered less rational than realists, as conformists conform to fraud because it is a frequent occurrence in society. Organizations looking to develop strong counter fraud efforts for this psychological profile would want to focus on the damaging effects that fraud has on society; the main message is that fraud is not socially acceptable and is quite damaging. This is where adding the perception of a victim may be helpful. This will illustrate that fraud is not victimless and not without damaging, broad-reaching impact. These individuals may be affected by more of a social approach to fraud fighting, and organizations that invest in strong public awareness and training may benefit from thwarting efforts from this specific fraudster profile.

Critics, or those that support and precipitate fraud, require the strongest and most direct counter fraud approach; organizations should focus on showcasing and communicating the presence of strong counter fraud detection systems and investigative procedures. Certain carriers I have worked with make it known to the public that they have highly trained fraud investigators and take a zero tolerance, aggressive approach to fraud fighting. This strategy will assist with helping to deter critics from engaging in fraudulent behavior in their respective companies.

As we have demonstrated above, there are many different psychological profiles of the fraudster, each with his/her own motivators, mind-set, moral compass, and triggers toward fraud. Each of these typologies will be affected differently by counter fraud efforts. Thus a savvy organization should utilize a multipronged approach at fraud prevention, an approach that serves to reduce opportunity.

Honesty and Cheating Behavior

I have noticed several intriguing academic studies on cheating and customer dishonesty that could offer certain insights in a counter fraud setting (Alm, Bruner, and McKee 2016). As discussed earlier, the relative lack of quantity of insurance fraud research has forced us to look into other areas for potential application; tax evasion and studies on cheating behavior serve this purpose. There apparently are many similarities in the behavioral aspects of fraudsters and those that cheat in other circles, and these are aspects that we can consider in our preventative efforts. There is a significant amount of research on cheating and lying behavior and how this is rooted in ethical decision making; one of my favorite researchers is Dan Ariely, from Duke University, who has published many articles and books (2013) on the topic.

Consider some of these alarming statistics:

- Thirty percent of Americans are honest all of the time.
- Forty percent are situationally honest.
- Thirty percent are dishonest.

What this reveals is that we, as fraud fighters, need preventative efforts that focus on 70 percent of the population. This challenge can be daunting when we consider the US population in 2016 was estimated to be 324 million people; that is 226 million people who are dishonest or situationally dishonest—all potential fraudsters!

Let's review additional alarming statistics from Ariely:

- Fifty-six percent of MBA students admit that they have cheated at least once, compared to 47 percent from other graduate programs.
- Forty-one percent of married couples admit to having an emotional or physical affair.

A significant amount of research (Ariely 2013; Xu and Ma 2015) that has been performed on cheating behavior can add insights into the motivational factors behind the fraudster. In-depth studies on cheating have revealed that participants that are presented with an opportunity to cheat for economic gain will do so if they feel they will not be detected, and there is a low penalty for the offense. Participants are more likely to cheat if they feel their chances of being caught and punished are minimal. It has been established that people are drawn to fraud due to the low punishment and high reward structure of the offense. From a pure prevention perspective, we should strive to publicize our successful fraud cases as much as possible in order to show that fraud is punished, which will hopefully push unethical individuals to consider their actions.

As the purpose of this book is to make a connection between psychology and fraud, we would be remiss if we did not address criminological theories and how they interact with cheating behavior. Two criminological theories could be applied in this topical area: social learning theory and self-control theory.

Social learning theory was first introduced by Albert Bandura in 1977 and refers to any social behavioristic approach to the social sciences. In technical terms, it focuses on the reciprocal interactions between cognitive, behavioral, and environmental determinants. This implies that there is a process between a stimulus and a response; in simple form, behavior is learned from observation (Bandura 1977). Ronald L. Akers and Edwin H. Sutherland further developed differential association as the most important aspect of the learning process; that is, behavior is learned from intimate groups, such as close family and friends (Lilly, Cullen and Ball 2015). This learned behavior is reinforced through the basic principles of operant conditions whereby rewards or punishment are provided as the result of actions. This is very useful to us in the fraud context, as the theory states that when any form of behavior is rewarded, then the definitions of that specific behavior are reinforced. Thus, it follows that those that receive rewards for deviant behavior (fraud) are more likely to develop positive definitions of that deviant behavior and will be more likely to commit it again. Social learning theory has received tremendous support through the years in many different research areas and is considered one of the strongest theories when attempting to explain criminal deviance, especially in the juvenile population.

Self-control theory was introduced by Travis Hirschi and Michael Gottfredson in the 1990s and operates under the hypothesis that individuals engage in deviant behavior because they cannot restrain themselves and delay gratification in any form (Akers 1991). There is a strong connection between control and age; as one ages, his/her self-control should also mature in a direct statistical relationship. We would expect adults to be able to delay gratification more than children, who often want instant gratification no matter the cost. Having a low degree of self-control has emerged as an extremely powerful theory to predict behavior and has been proven in many different criminal areas, such as drunk driving and theft.

Both of these theories have significant merit in the area of cheating and ethics. There could be certain similarities in behavioral and personality traits of fraudsters and those who have been the subject of studies on cheating and unethical behavior, thus making this research material of particular interest to us as we develop counter fraud efforts. Those who cheat in certain areas, such as on tests, taxes, shoplifting, and so forth, are also highly likely to cheat in other areas as well. This is of importance to us if we consider how we utilize background information in our fraud investigations. As fraud fighters, we should be very interested in those individuals that have been the subject of prior fraud investigations, have a prior criminal record, or have had other instances where we would believe they have been untruthful. This would be an ideal foundation for red flag indicators that would focus solely on these unethical behaviors.

An international fraud case that occurred near my hometown is worth discussing. The fatal *Ethan Allen* tour boat tragedy of 2005 is a great case that exemplifies many of these points discussed. In this tragic accident, a boat carrying tourists overturned and sank in Lake George, New York, killing twenty people. What began as a local investigation resulted in an international web of international money laundering, tax evasion, and insurance fraud. The story begins in 2003 when Robert Mills and Chris Purser founded the Global Property Owners Association, a Florida-based insurance company. Chris Pursuer had previously been restricted from performing any insurance transactions in Texas as a result of a fraudulent activity. In the spring of 2004, James Quirk, president of Shoreline Cruises, purchased a $2 million insurance policy from Global Partners covering the tour boat *Ethan Allen*.

Global Property is provided with false financial support documents for the policy by Malchus Irvin Boncamper, a Caribbean-based individual who created significant capital committing money laundering and tax evasion. In October 2005, the *Ethan Allen* overturned in Lake George, New York, while engaging in a routine tour, killing twenty passengers. When Quirk contacted Global Properties to report the claim, he was immediately advised that he had no coverage, and thus an investigation ensued, managed by various federal and state agencies, including the FBI and the IRS. In 2011, a federal grand jury indicted Purser, Mills, Boncamper, and several others, who pleaded guilty to money laundering, wire fraud, and securities fraud in relation to the insurance fraud scheme.

This case illustrates that there is significant overlap in many areas of white-collar crime and that fraudsters are often involved in other areas of deviant, cheating behavior. This showcases the importance of looking very closely at claims where involved parties have multiple prior incidents or have had prior involvement or connections with other parallel crimes such as tax evasion or money laundering.

In recent years I have seen a growing body of research pertaining to honesty statements and their credibility in many different areas. Thus, it is worth discussing in an insurance fraud context. We have come to an understanding that there needs to be different preventative approaches geared toward the opportunistic and organized fraudster, as each group is characterized by its own distinct behavioral and motivational traits. When we speak of triggering internal honesty, we have to assume that the individual is capable of having his or her internal honesty triggered. In other words, we are assuming that these people are rational thinkers (recall these principles from earlier chapters) and that their behavior would be modified by using these honesty declarations. To a hard-core, organized fraudster, who is attempting to orchestrate a complex staged accident ring, an honesty statement may not serve any deterrent purpose. But, to an opportunistic fraudster that may be already feeling even a slight degree of anxiety over his or her filing of a fraudulent claim, an honesty statement may have an impact.

Triggering internal ethics and morality has been the focus of many research studies in recent years, such as a very interesting study performed in an insurance fraud context by Leal et al. and Mann (2016). These studies focus on methods that trigger an individual to *check in* to his or her moral standards and question the specific action he or she is engaging in. This is becoming a common practice in academics, where we may ask a student to sign an honesty declaration at the start of a course, asking them to be truthful in their homework submissions, avoiding plagiarism and cheating on tests, and so on. These statements are usually incorporated into a course syllabus; however, I and several other instructors also insert them into specific homework and testing activities, hoping to trigger additional honesty at multiple times during a course. Research in applying these declarations in the academic arena is very positive, showing that they do in fact have an impact on the ethical submission of class work. In one school I taught for, its honesty protocol was to require the students to attach the declaration to each and every activity. Here is a sample we used:

School of Criminal Justice—Honesty Declaration
Homework Activity—Week 3

This honesty declaration form must be completed in its entirety and attached to your completed homework activity for week 3. Only activities that have this declaration attached will be accepted for grading by your instructor. If you submit an activity without this declaration, then you will receive a zero "0" for this week's activity.

I, the undersigned, declare that the attached homework activity for week three is the product of my own work, and that no part of it has been plagiarized or copied from any source. In addition, all of the information contained herein is truthful to my knowledge and had not been altered in any manner.

First and Last Name: _____

Student ID Number: _____

Course/Section: _____

Assignment: _____

Date: _____

Signature:
_____

And here is one I utilized in another school, but only at the beginning of the course.

ACADEMIC-HONESTY CONTRACT
Dr. J. Michael Skiba
CRJ555-Fraud Data Analytics

I will do my own original work. I will not copy the work of another person, in whole or in any part.

I will not receive unfair/unnecessary assistance from another student, parent, computer program, or any other unauthorized source on a project, or assignment that is meant to be completed alone.

I will not consult other unauthorized material or information during tests unless instructed by my teacher (calculator, electronic storage, notes, etc.).

I will not plagiarize. I understand that plagiarism is using the words or ideas of other authors and artists in my work without giving the authors credit.

I will not take material from the Internet or another student's electronic files and use it as my own. I will not copy text, graphics, musical scores, mathematics solutions, presentations, or any idea in any form from another source without proper citation.

I will not communicate exam information or answers during or following an exam.

I will not claim credit for work that is not the product on my own honest effort.

I will not turn in an original paper or project more than once for different classes or assignments.

Any student who breaches this Academic Honesty Contract is subject to disciplinary action that includes receiving NO credit for homework, project, etc.

In the event of testing, the student will be required to take a different version of the test for credit.

I have read and understood this Academic Honesty Contract. I will follow the rules stated above.

Student Name: (Print) _____

Student ID Number: _____

Date: _____

Signature: _____

I have also taught online collegiate courses, which present extremely challenging ethical dilemmas, as the assessments, quizzes, papers, and virtually all activities are completed in a totally unsupervised, unstructured format. In this environment, honesty issues are a constant concern for instructors facilitating those courses. Statistics reveal that almost 90 percent of all higher education institutions are offering virtual courses, implying that this will be a trending challenge among universities.

In my opinion, as a professor for over ten years, the degree of supervision required in an online format depends entirely on the level of class being facilitated. Of course, no matter the level of the class, there is always the potential for student dishonesty; however, I have noted that the introductory-level courses, such as Introduction to Criminal Justice or Introduction to Business Management, appear to have more instances of dishonesty than the upper-level courses such as Cybercrime or Terrorism Investigations. The student demographic of these courses varies greatly, with the intro-level courses having students in the early stages of their career, or not employed at all, and the upper-level courses having students that are in professional, mid- to upper-level positions looking to further their careers. This would be an interesting area for further study on a personal level.

How might you ask are methods that professors can use to reduce dishonesty in the virtual classroom? Seminal studies have been performed in this area that are highly relevant and offer us suggestions. Some of these tactics, although not directly applicable in a fraud setting, can offer suggestions and a theoretical basis for strategic prevention in our fraud industry. Communication and constant reminders regarding dishonesty is seen as a very effective approach; e-mails, discussions, and posting the policy in many locations is highly suggested. Professors can show the actual academic honesty policy, post warnings that list the consequences and links to the policy can be offered along with reminders of how an honesty violation can affect students on a long-term professional basis. A second method is to require the students to submit their work electronically and send it through a plagiarism checking system; many are commercially available. A third strategy is to closely manage the exam-taking procedure, including strict time limits and rotating questions among students. All these approaches have proven to be very effective at reducing incidents of online academic dishonesty.

Current research under way focuses on the location of the honesty declaration and whether the placement of these in any particular location has any positive or negative effect on triggering ethical morality. Interestingly, initial findings revealed that when the honesty declaration was placed at the very beginning of a document, then this triggered increased morality and had a more significant impact. A colleague who works for an international consulting group is the primary investigator of a research project in this area; the study is not complete, but he advises that his initial results show an approximate 10 to 20 percent decrease in unethical responses when the statement is placed at the beginning of a document as opposed to being placed at the very end. Seminal research performed by Shu et al. (2012) also support this finding. Their research study revealed that placing an honesty declaration at the beginning as opposed to the end reduced unethical replies by 10.25 percent, an extremely significant result.

Other areas of criminal justice have incorporated this principle; when a witness testifies in court, we ask him or her to be truthful prior to the testimony, not at the conclusion. This declaration can become extremely valuable at the point of sale, as we often require insureds to self-report policy information such as driver's age, drivers in household, residency, and the like, when it is financially advantageous for the insured to misrepresent information. Adding a simple statement at the very beginning of an insurance application, before he/she starts to complete it appears to be an effective method to reduce unethical responses. These honesty declarations can be used in areas of insurance, including policy applications, claim forms, statements, EUOs, receipts for claimed items, proof of loss forms, first notice of loss reports, and many other documents.

In the United Kingdom, researchers studied the placement of the declaration and also found merit in including this at the very beginning of a document as opposed to the conclusion (Cabinet Office Behavioural Insights Team 2012). Government agencies decided to adopt this perspective and have incorporated placing these honesty statements at the beginning of many documents in their operations, including on government tax return forms. They found that when the declaration is placed at the end of the document, very rarely did the participant go back and review and modify his or her initial entries. Furthermore, there appears to be no preference among those completing these forms of placement at the beginning or at the end, so it therefore behooves an insurance company to consider placing them at the very beginning.

I have consulted with several carriers and agencies that have experimented with this declaration and have seen very positive results. A full quantitative analysis was not performed, so there is no published data to share, but my observations and their feedback revealed that this approach has definite merit. So what should be included in an honesty declaration? Studies show that in order to trigger morality, the statement can be very short and simple, but the word *honesty* should be a component.

A simple declaration could be as follows:

Honesty Declaration: "I hereby affirm that the information provided in this policy application is an honest and true representation to the best of my knowledge."

Several carriers incorporate a zero tolerance fraud statement into the honesty statement, which is a highly effective and recommended approach for multiple reasons. We have spoken earlier about how showcasing zero fraud tolerance is highly suggested as a fraud strategy, as it helps fraudsters to not view you as a soft target. If this zero tolerance statement is coupled with an honesty statement, then this is an incredible multipronged approach toward fraud fighting, as it will have a deterrent effect (discussed earlier) and also trigger internal honesty. It is extremely important to work with one's legal department in developing these statements to ensure compliance with corporate policy.

Thus far in our analysis of honesty and cheating, we have discussed the relative applicability of studies in this area from an external perspective. But we cannot forget the internal perspective, as employee theft and fraud is on the rise. Fraud departments must be able to counter this problem, as studies reveal that this is an area of mounting concern and financial exposure for companies. In my career, I have seen an increasing interest in fraud detection that focuses on internal theft; this interest is driven by thefts that they themselves have experienced within their respective companies. What seems to come as a surprise to these companies is how the particular employee was able to carry out his or her criminal actions.

Consider these disturbing statistics as published by the Association of Certified Fraud Examiners (2016) Global Fraud Study:

- Five percent of corporate revenues are lost annually to internal theft.
- Estimated loss is $3.5 trillion nationwide.
- Average fraud loss is $140,000 per instance.

- More than 20 percent of cases have over $1 million in losses.
- Frauds last approximately eighteen months before detection.
- Industries most victimized by internal fraud are banking and financial services (insurance companies).
- Presence of internal anti-fraud controls shows significant decline in internal fraud cases.
- Most internal fraudsters are first-time offenders with clean employment records.
- Eighty-seven percent of internal offenders have never been charged with a fraud-related offense.
- Eighty-four percent have never been punished or terminated by an employer.

One international carrier I worked with had experienced an internal theft orchestrated by one employee over two years that totaled $2 million. This carrier had certain internal fraud countermeasures in place, but because this employee had access to all the programs, he was able to circumvent all of these systems and act undetected. I recall two other cases of internal theft to which I was exposed. In both of these instances, the employee conducted his criminal activities using the access he had to certain payment systems; one was a claim settlement program, and the other was a vendor payment system. In the first case, the employee was fabricating claimants on legitimate claims and issuing payments to himself under those names. All the claim payments were under his claim authority of $10,000, so they did not require any approval or anyone else to review the check. In the second instance, the employee would make payments to fake vendors on claims and then cash the fake vendor checks. Again, those checks were all under this employee's threshold amount of $5,000. As we can see, the main commonality among these cases is that the carriers did not have any formal system of checks and balances in place, a system that is very simple to implement.

In the broader white-collar crime landscape, there are five main categories of employee theft, also called occupational or internal theft or fraud:

1. Theft of information, such as client lists and other trade secrets, for gain on a personal or professional level.
2. Embezzlement/misappropriation of funds.
3. Skimming, when funds are diverted or when cash is skimmed off the top.
4. Fraudulent disbursement, such as when expense reports are inflated.
5. Larceny-theft.

In the carriers I have worked with, most of their exposure came from #5—outright larceny. This appears to be the area that companies should focus on as a portion of their internal fraud strategy. When we insert the theories of criminology we have been discussing throughout this book into the analysis of internal theft, most internal offenders would be categorized as opportunistic criminals. Most employees that commit theft have no prior records, are good-performing employees, and are solid, upstanding citizens, yet they engage in criminal deviant behavior.

As a central theme of this book, reducing opportunity should be the primary focus of carriers in developing anti-fraud strategies, and approaching internal fraud is no exception. If we apply tenets of rational choice theory into our analysis of the internal fraudster, we can implement programs that will serve to deter the rational-thinking employee. Many of the strategies we have been discussing in regard to external fraud also apply to internal thefts. Some examples would be showcasing other cases that were investigated/prosecuted, passing and posting strict zero tolerance statements, and requiring honesty declarations.

The use of technology also can assist companies to flush out internal thefts before they occur, providing significant savings. Technology can be used to track internal payments and to red flag administrators when certain ones are triggered. Specific red flags can be developed and integrated into software programs to track the behavior of an employee pertaining to check transactions, amounts, vendors, payees, and so forth. In the earlier example I provided of the case that cost one carrier $2 million, the company decided to implement a software system to help identify and filter internal fraudulent activity. After the initial consultation and vulnerability assessment was performed, a list of red flags was developed, which was used to structure the rule-based detection system. Software systems can detect frequent transactions based on dollar amounts, addresses, payees, time frame, vendors, and many more.

It is vital that companies develop certain strategies to help thwart internal theft issues, and I have found that most companies and agencies are satisfied in acting in a pure reactive manner when these situations arise. In this section, we will discuss several alternative strategies that have proven credible. In the earlier examples of fraud, I mentioned a system of checks and balances; let's expand on this strategy.

Checks and Balances

Using a system of checks and balances is beneficial, as it ensures that no single person has control over all aspects of a financial transaction, this is also referred to as the *four eyes principle*. In an insurance scenario, this could involve a tiered system of financial threshold; claim rep A has to have all checks approved by supervisor B, and supervisor B's checks are approved by manager C, and so on. Under this structure, there are always two individuals looking at every single transaction, which would prevent a single individual from being able to issue checks completely autonomously. Another strategy would be to require one person to write the check and another to sign it, which again would imply two reviews of a transaction.

Leveraging technology can also be an extremely effective method to insert into a company's system of checks and balances. Internal and external technology can be used to help identify patterns and outliers in the issuance of checks within a company. I have seen solutions whereby the system would track patterns in regard to the issuance of checks, such as an employee that issues many checks for similar amounts, to the same vendor, on the same claim. It is also important to look not only for repetition but also for rare outliers in the data. For example, if an employee issues most checks under $2,000, but issued one for $10,000, this could easily be flagged and identified using technology, prompting additional investigation into the matter.

In many companies, no matter the size, employees often have many responsibilities and accountabilities, more so now than in years past. The danger from an internal fraud perspective lies in that one individual may be responsible for reviewing checks, issuing checks, reviewing invoices, and making deposits, all without any additional oversight, which results in a significant vulnerability to the company. One of the most important aspects of a successful system of checks and balances lies in effectively dividing the incoming and outgoing financial tasks. In other words, at a very minimum, a company should implement a system whereby different employees handle the incoming cash, checks, and invoices, and a different employee handles the outgoing payments and products.

Vacations?

Surprisingly, a large number of internal fraud cases are discovered when the offender takes a vacation! When an employee is on vacation, others often have to delve into areas of the employee's specific accountabilities, which often reveal fraudulent activity. Vacations are a forced system of checks and balances. There are documented cases that demonstrate that internal fraudsters often take fewer vacations than others, as they feel that their fraud has a greater chance of being discovered if they are absent. Oftentimes internal fraud is uncovered by a simple situation that needs verification by the vacationing employee, yet in their absence, discrepancies are noted. I recall one situation when a vendor called to have a check reissued; the assigned claim employee was out on vacation, and thus a colleague had to delve into the file and reissue the check. The colleague discovered that the claim representative and vendor had an unethical arrangement in their vendor relationship. All of this would not have been discovered if the employee was not on vacation. Companies should be wary of employees who carry over a large number of vacation days per year without any reasonable explanation.

Code of Conduct

A clearly posted code of conduct is an essential component toward developing a strong internal anti-fraud program. Employees must have a clear understanding of where the line is between normal business activity and that which the company deems unethical and in violation of the business code. I perform a significant amount of corporate ethics training, and a lot of companies had no code of conduct at all. Not only was it not clearly posted, but it did not exist! This is a clear recipe for fraud, as it allows complete latitude in business dealings and provides no parameters for employee conduct. Other companies have extremely strict codes of conduct, whereby employees are forbidden from accepting anything (even pens) from vendors or other outsourced providers. Other companies have lengthy codes, sixty pages or more, and outline everything from workplace safety to disability procedures.

Statistics reveal that about 89 percent of organizations have a formal ethics program, which is a great result; however, about half of the employees within these organizations state that these programs are unclear and misunderstood. Even more disturbing is the fact that of the 89 percent of organizations that have a formal program, only 46 percent are supported by training and communication (Association of Certified Fraud Examiners 2016). This means that almost half of all employees within an organization are totally unclear and unsure of the ethics protocol, program, and parameters—an incredibly troubling statistic!

If a company is looking to develop a strong ethics program, the first step is to create a code of conduct that fits into the corporate culture; second, it must be posted and visible to employees; and lastly, awareness training must be provided. The code should clearly identify unacceptable behavior and then the repercussions to employees who violate it. If it is not feasible to post a lengthy code of conduct, the company can create a brief statement, similar to a mission statement, that succinctly states the code. This should also be provided to new employees, who should be asked to read, sign, and consent to the code. This written consent should also be readministered on an annual basis. The signed code of conduct agreement has a similar effect as honesty declarations (as we discussed at length earlier in the book) and serves to trigger internal ethical behavior.

The following is an example of a sample code of conduct as presented by the Association of Certified Fraud Examiners (2016).

Sample Fraud Policy

Background

The corporate fraud policy is established to facilitate the development of controls that will aid in the detection and prevention of fraud against ABC Corporation. It is the intent of ABC Corporation to promote consistent organizational behavior by providing guidelines and assigning responsibility for the development of controls and conduct of investigations.

Scope of Policy

This policy applies to any irregularity, or suspected irregularity, involving employees as well as shareholders, consultants, vendors, contractors, outside agencies doing business with employees of such agencies, and/or any other parties with a business relationship with ABC Corporation (also called the company).

Any investigative activity required will be conducted without regard to the suspected wrongdoer's length of service, position/title, or relationship to the company.

Policy

Management is responsible for the detection and prevention of fraud, misappropriations, and other irregularities. Fraud is defined as the intentional, false representation, or concealment of a material fact for the purpose of inducing another to act upon it to his or her injury. Each member of the management team will be familiar with the types of improprieties that might occur within his or her area of responsibility, and be alert for any indication of irregularity.

Any irregularity that is detected or suspected must be reported immediately to the Director of _____, who coordinates all investigations with the Legal Department and other affected areas, both internal and external.

Actions Constituting Fraud

The terms *defalcation, misappropriation,* and *other fiscal irregularities* refer to, but are not limited to, the following:

- any dishonest or fraudulent act
- misappropriation of funds, securities, supplies, or other assets

- impropriety in the handling or reporting of money or financial transactions
- profiteering as a result of insider knowledge of company activities
- disclosing confidential and proprietary information to outside parties
- disclosing to other persons securities activities engaged in or contemplated by the company
- accepting or seeking anything of material value from contractors, vendors, or persons providing services/materials to the company with the exception of gifts less than fifty dollars in value
- destruction, removal, or inappropriate use of records, furniture, fixtures, and equipment
- any similar or related irregularity

Other Irregularities

Irregularities concerning an employee's moral, ethical, or behavioral conduct should be resolved by departmental management and the Employee Relations Unit of Human Resources rather than the _____ Unit.

If there is any question as to whether an action constitutes fraud, contact the director of _____ for guidance.

Investigation Responsibilities

The _____ Unit has the primary responsibility for the investigation of all suspected fraudulent acts as defined in the policy. If the investigation substantiates that fraudulent activities have occurred, the _____ Unit will issue reports to appropriate designated personnel and, if appropriate, to the Board of Directors through the Audit Committee.

Decisions to prosecute or refer the examination results to the appropriate law enforcement and/or regulatory agencies for independent investigation will be made in conjunction with legal counsel and senior management, as will final decisions on disposition of the case.

Confidentiality

The _____ Unit treats all information received confidentially. Any employee who suspects dishonest or fraudulent activity will notify the _____ Unit immediately, and should not attempt to personally conduct investigations or interviews/interrogations related to any suspected fraudulent act (see Reporting Procedure section below).

Investigation results will not be disclosed or discussed with anyone other than those who have a legitimate need to know. This is important in order to avoid damaging the reputations of persons suspected but subsequently found innocent of wrongful conduct and to protect the company from potential civil liability.

Authorization for Investigating Suspected Fraud

Members of the Investigation Unit will have

- free and unrestricted access to all company records and premises, whether owned or rented, and
- the authority to examine, copy, and/or remove all or any portion of the contents of files, desks, cabinets, and other storage facilities on the premises without prior knowledge or consent of any individual who might use or have custody of any such items or facilities when it is within the scope of their investigation.

Reporting Procedures

Great care must be taken in the investigation of suspected improprieties or irregularities so as to avoid mistaken accusations or alerting suspected individuals that an investigation is under way.

An employee who discovers or suspects fraudulent activity will contact the _____ Unit immediately. The employee or other complainant may remain anonymous. All inquiries concerning the activity under investigation from the suspected individual, his or her attorney or representative, or any other inquirer should be directed to the Investigations Unit or the Legal Department. No information concerning the status of an investigation will be given out. The proper response to any inquiries is: "I am not at liberty to discuss this matter." Under no circumstances should any reference be made to "the allegation," "the crime," "the fraud," "the forgery," "the misappropriation," or any other specific reference.

The reporting individual should be informed of the following:

- Do not contact the suspect individual in an effort to determine facts or demand restitution.
- Do not discuss the case, facts, suspicions, or allegations with anyone unless specifically asked to do so by the Legal Department or _____ Unit.

Termination

If an investigation results in a recommendation to terminate an individual, the recommendation will be reviewed for approval by the designated representatives from Human Resources and the Legal Department and, if necessary, by outside counsel, before any such action is taken. The _____ Unit does not have the authority to terminate an employee. The decision to terminate an employee is made by the employee's management. Should the _____ Unit believe the management decision inappropriate for the facts presented, the facts will be presented to executive level management for a decision.

Administration

The director of _____ is responsible for the administration, revision, interpretation, and application of this policy. The policy will be reviewed annually and revised as needed.

Approval

_____
(CEO/Senior Vice President/Executive

There is one significant word mentioned in this statement that bears further discussion: *reporting*. Just as a company should focus on the identification of violations of the code, it should also be concerned with the reporting process, and accordingly this should also be included in the code. It should be mentioned and stressed that reporting violations of the code is a very serious matter and one where nonreporting is also a serious offense with repercussions.

Nonprofit organizations should also create a strong and powerful Code of Ethics, such as this succinct code, which belongs to the International Association of Special Investigation Units (2016).

1. Members will, at all times, act with integrity, trustworthiness, and honesty, modeling the professional behavior exemplary of our organization.
2. Members will diligently and competently assume and perform their professional responsibilities.
3. Members will respect the sensitive nature of any confidential or proprietary information made known or available to them, consistent with the law. Members also will exercise the care required to prevent the unlawful or improper disclosure of such information.
4. Members will take the appropriate care to protect or maintain the reputation and professional practice of a colleague, employer, insured, claimant, or other subject of an investigation.

5. Members will demonstrate the requisite courage and responsibility in their decision making when faced with potential conflicts of interest.
6. Members will perform their duties and, at all times, conduct themselves in accordance with the law.
7. Members will, at all times, be aware of and adhere to the principles of appropriate antitrust laws.

One last critical area of the code of conduct pertains to the activity that ensues after a violation is reported. It is very important for employees to see strict follow-up, and enforcement, of any code violations. Swift and severe punishment are the main tenets of classical criminology, as discussed in the criminological theory section of this book. Studies have revealed to us that a swift and severe response to code violations will serve to create a deterrent effect, which will help reduce overall internal fraud occurrences. A sample policy statement for this follow-up would be as follows:

Due to the important yet sensitive nature of the suspected violations, effective professional follow-up is critical. Managers, while appropriately concerned about "getting to the bottom" of such issues, should not in any circumstances perform any investigative or other follow-up steps on their own. Concerned but uninformed managers represent one of the greatest threats to proper incident handling. All relevant matters, including suspected but unproved matters, should be referred immediately to those with follow-up responsibility.

And finally, the written declaration:
Approval

_____   _____
(CEO or other designated executive)   Date

================================

Acknowledgment

My signature signifies that I have read this policy and that I understand my responsibilities related to the prevention, detection, and reporting of suspected misconduct and dishonesty.

I further acknowledge that I am not aware of any activity that would require disclosure under this or other existing company policy or procedure statements.

Signature: _____

Print Name: _____

Date signed: _____

As we have demonstrated, a clear code of conduct, and one that is visibly posted, communicated, and signed by employees, is one additional tactic companies can utilize as part of their multipronged approach to fighting fraud. These approaches have been proven to generate positive results in helping to thwart instances of internal theft, an increasing concern in today's corporate environment. As mentioned, employees must operate in an environment where fraud and ethics are made a priority as part of business dealings; this corporate culture needs to incorporate ethical principles to be truly effective. It is critical for employees to see others in the company complying with the code; if a clear code is posted, but upper management is seen violating the policy, then there is no doubt that subordinate employees will also take advantage of the code. Thus, at all levels of the company, a positive corporate culture must be created and precipitated.

One of the most important, and often overlooked, steps in this process is offering ethics awareness training. As we learned earlier, almost half of all businesses do not offer training and do not communicate these important ethical issues. I have performed hundreds of ethical training sessions, which ranged from multiple-day workshops to simple one-hour meetings. Remember—something is better than nothing. I have witnessed great results in companies that make ethics training a part of an annual or biannual corporate meeting. As a presenter of corporate ethics training and professor of ethics classes, I can offer that this topic is quite fun and can be very engaging for an audience. A role play format and heavy audience involvement in various ethics scenarios are great tactics to utilize in these training and awareness courses.

Culture of Honesty

Creating an ethical company starts from the top down; the CEO and leadership must make ethics an ingrained part of the work environment. The actions of upper management must also reflect the importance of ethics. They must act in compliance with all policies, especially if they desire for others in the company to adhere to those same standards. Corporate leaders should focus on ethics for morality reasons, but the evidence shows that companies that make this a high priority are also more profitable. It is widely known, and documented by many sources, such as *Forbes Magazine* and the Association of Certified Fraud Examiners, that companies that encouraged open communication had a well-defined code of conduct, and placed high priority on ethics show greater profitability. In addition, companies that promote open communication have employees that are happier and more satisfied in their roles, all which lead to increased work product. The word *transparency* has become a popular term used to describe many areas within a company and implies that there are no issues that are off-limits to discuss.

What are some tactics that a company can implement to help create this culture of honesty? First, firms need to focus on the hiring process, as will be discussed further in a later section. Starting with ethical employees is the key to maintaining an ethical work environment. Second, there needs to be a positive work environment; a company has to develop and publish a clear code of conduct that creates the parameters for behavior. A third strategy is to provide employees with formal assistance programs. These programs serve to assist employees with personal situations, family issues, or financial stressors, basically offering a line of communication and formal assistance to employees that might otherwise fall into a high-risk category for fraudulent activity. These programs help to reduce the pressure to commit fraud should an employee be presented with some financial and/or personal stressors.

A final strategy would be to promote that employees partake in ethics training classes, whether they are off-site at conferences, on-site in a large group format, on-site in smaller group settings, online, or any other format. The research supports the efficacy of ethics training classes in reducing unethical behavior (Weber 2015). I am surprised as to how many companies do not engage in this very simple yet effective method to increase ethical awareness. I am also surprised as to how many collegiate programs lack an ethics course or other morals class as part of their curriculum. Introducing ethical concepts at this academic level could be a highly effective way to engrain ethical principles at a very early and important stage of professional development. I have seen very rigorous MBA programs that only contain ethics as part of a small segment (maybe one or two weeks) of a twelve- or fifteen-week management class; I rarely see ethics as an entirely separate course offering, which is very unfortunate.

It is important for those supervising ethics programs within a corporation or business to remember that no matter what format is used, the training should be seen as applicable to the employees' actual work environment and not viewed as a generic training session. I recall one on-site ethics training session I performed and remember overhearing the attendees remarking how they perceived the training course's main purpose was to reduce liability on the company should an ethical situation arise. In other words, the employees felt the training was part of a cover yourself mentality coming from the company and was not intended to actually help the employees recognize and manage ethical situations. All ethics training will have generic principles, but I have received great feedback in training and courses I have facilitated that integrated real-world situations that were similar to those particular employees' work scenarios. My experience also shows that using group discussions in these training sessions also helps to strongly influence an individual's integrity using this collective influential approach. This helps to create a more in-depth awareness of ethical dilemmas and focuses on how these employees would handle these situationally based scenarios.

Hiring Practices

As already mentioned, hiring moral, ethically sound employees is one of the first, and most effective, means to create a culture of honesty. In recent years, I have seen companies place an increasing emphasis on spending more resources on the hiring process; this is incredibly beneficial, as investing time and resources at this early stage has shown to offer a great return on investment. Studies show that approximately 25 percent of most internal theft cases are conducted by employees who have worked for the company three years or less, further supporting the need for focus at this early stage.

The following are some specific tactics companies can utilize in hiring ethical employees.

- Clearly mention strong ethical standards as part of the initial job announcement; this will set the stage very early on that ethics plays a major role in the corporate culture and may deter unethical individuals from applying.
- Once the resumes have been received, use manual reviews or computer systems to identify red flags of potential ethical issues, such as gaps in employment, frequent job changes, or potential disciplinary actions.
- Once stronger candidates have been identified, ask for them to complete an application, which can then be compared to the previously submitted resume for potential discrepancies.
- If face-to-face interviews are scheduled, ask the candidates to complete an honesty declaration, which will also reinforce the strong ethical culture of the company.

- Administer an honesty test, such as a Personnel Selection Inventory (PSI), Reid Report Risk Assessment, or a Veracity Analysis Questionnaire (VAQ).
- During the face-to-face interview, ask behavioral situational questions such as "Have you ever witnessed a fellow employee stealing" or "Have you ever stolen anything in your life, even say a pen from a fellow student?"
- Once these formalities are complete, then it is critical to perform a detailed background check on the remaining pool of candidates. At this stage, it is important to check with all references, prior companies, prior supervisors, colleagues, and schools and look for any unethical flags that could indicate a future problem.
- As a final step, performing a detailed credit check will reveal any prior financial situations that could also create pressure as part of our fraud triangle motivators. Prior credit problems usually translate into future credit/financial issues.
- Once the employee is hired, it is also very important to be diligent and continue with the process. Performing an employee audit is significant, as it will serve to locate any gaps or other potential issues that were missed by the initial testing.

If you are thinking that this process seems long and daunting, you are correct. However, companies that spend resources at the hiring stage have proven to have fewer incidents of internal theft and to have increased profitability. These companies report a very significant return on investment when devoting resources to the hiring process.

There are many different methods that companies can use to hire employees, and we have discussed some of the most effective ones thus far. However, the research shows that the actual face-to-face, conversational, situationally based interview format is the strongest measurement of employee integrity, with most hiring managers reporting they believe this is the most effective approach. Managers should be able to perform a successful face-to-face ethically based interview by asking situationally based questions and then listening very closely to the responses. Any response that may indicate a lack of integrity, dependability, accountability, or work ethic should be probed for deeper understanding by the interviewer. Even trivial responses that reveal a lack of employer policies such as sick time, absenteeism, tardiness, performance issues, and so on, all could indicate an integrity issue that could develop into a significant issue later.

The following questions are just a sample that could be used during an interview that helps to assess a candidate's ethical compass; these questions have been proven very credible and reliable at filtering ethical employees. Remember—listen closely to the responses!

- In your opinion, what makes a company ethical? This is a very broad question, but one where the interviewee has a lot of flexibility in his/her reply. It is important to listen carefully to the response, and then follow-up questions can be tailored accordingly. Did the candidate mention a specific area of a company such as sales? Did he or she discuss a code of conduct? This broad question can lead to many follow-up questions that can help hone in on the candidate's ethical makeup.
- How do you feel about a code of conduct, and have you worked for a company that has a well-defined one? This question, similar to the first one, is global in scope and will help to reveal certain areas for further exploration. Follow-up questions could be focused on if he/she witnessed anyone violating this code, and on what the result was, what the interviewee did, and so on.
- Tell me about a time that you faced an ethical challenge. Be very wary of a candidate that claims he or she has never faced an ethical dilemma; this usually indicates that the person may have been in fact directly involved in an unethical situation. Definitely probe further if this response is received. The most important aspect to keep in mind here is that the candidate is open to discussing this with you; we would rather have a known integrity issue than an unknown one.

- When you had the ethical situation above, who did you talk to? Again, the important perspective to keep in mind here is that the employee is maintaining an open communication channel with the company and not keeping the issue to him/herself. We want an employee that has consulted with a fellow employee, an HR manager, and so forth. Be concerned with those candidates that did not communicate the scenario to anyone.
- Have you ever had a conflict with your supervisor? Yes, we all *love* this question! As an interviewer, keep in mind that the red flags here pertain to a lack of communication—that is, an employee who maintains that he or she had a conflict with his or her manager and resolved it without consulting with him/her. Be suspect of employees that respond with a firm affirmation that they have never had a conflict; again, this usually indicates that they have but don't want to reveal this to you.

Companies should place emphasis on the hiring process, utilizing the different methods discussed briefly here. This should be a priority, as internal theft is becoming an increasing concern in our industry, with more companies looking for solutions to this in-house threat.

Internal Training

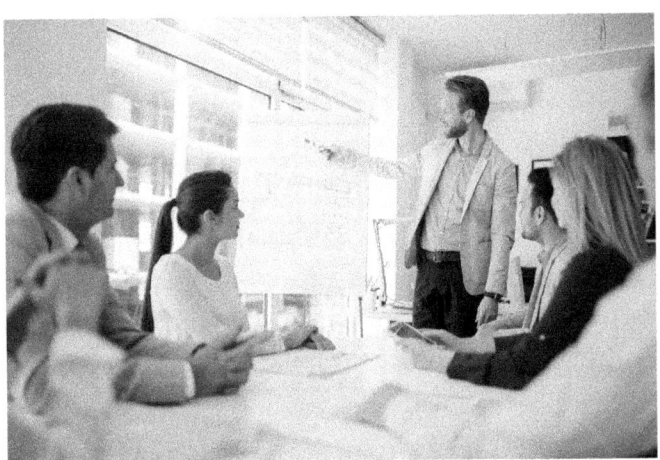

We have established that hiring ethical employees should be a significant priority for companies, as having these key people in the right places will make a more robust and effective counter fraud program. However, we also need to make sure all of our internal staff is well trained and fully informed of the fraud problem. Conducting internal training in the form of: sessions, presentations, workshops, and kick-off meetings is seen as one of the most effective methods to increase the effectiveness of a fraud strategy. Conducting training accomplishes two key tasks: first, it provides tangible data and information to the staff on the problem and where the current and potential vulnerabilities are; second, it creates a personal connection between those in the actual training session. This connection is highly overlooked yet is an extremely important aspect of a successful fraud program, as communication across departments, units, and disciplines is critical for highly focused strategies. In one carrier I routinely perform training for, there was a rolling joke pertaining to how its fraud referral rate increased significantly after I performed a training session. I recall one specific training session that involved both the fraud and front line claims unit; the topic of the session was focused on performing a successful interview as part of an investigative tool. The session ended on a Monday, and there was a remarkable spike in fraud referrals starting the very next day! Again, the training was successful because it increased the awareness of the fraud problem and served as a reminder to the internal staff that fraud should be a priority.

While working in the Nordics, I recall a very interesting conversation I had with a corporate security executive for a large financial services company. His organization became the victim of cybercrime. A malicious virus was inserted in its internal computer systems, which cleaned out about one-third of its files before the organization could counter it. After this incredibly costly and damaging situation, he embarked on an extensive internal penetration test to assess the status of his internal security. He sent out two hundred fake malicious test e-mails and physically mailed fifty USB sticks to several offices in unmarked envelopes. About 20 percent of the employees either clicked on the malicious e-mail or inserted the USB stick into their computers, a very alarming percentage. Considering it takes on average 164 days to detect ransomware, it is critical that the internal staff is well trained and well educated as to the damaging effects of this crime.

Almost all the carriers and companies that I have worked with through my career have remarked that offering training is an extremely worthy investment of time and resources. Training serves to formally educate but also simply create an awareness of the fraud problem. The resistance to perform training often comes in the form of the lack of ability to quantify its worthiness. However, I would argue that all other factors being equal, companies should monitor their fraud rates, and I would guarantee that they will see an increase in fraud referrals following any form of training session.

Wages and Internal Theft—Entitlement Theory

Many fraudsters commit fraud as they feel a sense of entitlement; they justify their activity in a manner that makes them feel *better* about the crime. Internal employee fraud is no different. Many fraudsters commit fraud out of retaliation toward some aspect of their working conditions, which often includes compensation. I witnessed one company go into complete turmoil when, because it pledged complete transparency, decided to release corporate salaries. The company did this in a manner that did not reveal the names of specific employees but instead the particular compensation of a salary band. This was retracted within a year, as it caused incredible tension, turmoil, and negative energy among the employees.

Academic studies have also confirmed this contention. Wage disclosure can cause increased incidents of employee fraud, as it breeds a sense of entitlement with the employee. One particular study by Chen and Sandino (2012) revealed that when a salary gap is identified, the employee at the lower salary range will oftentimes act negatively in some manner, not necessarily in a fraudulent manner, but in a way that has negative connotations. This behavior is justified by the lower-paid employee as an entitlement to recoup the wage gap; he/she will feel justified in the actions as he or she feels undervalued by the company. Thus, there appears to be a threshold on the transparency that a company should show, and revealing compensation has been demonstrated to result in negative outcomes.

In the above scenario of wage disclosure, it is difficult in the insurance and financial sector industry to accurately quantify fraud due to the complexity of measuring these incidents. In the Chen and Sandino (2012) study, the research revealed that higher wages have a positive impact on employee honesty. In one midsize retail store chain, researchers discovered that increasing salaries by $1 would cost $16,285, but prevent $6,362 in theft. A return on investment calculation would reveal that the wage increase of $1 would only cover 39 percent of the cost; however, this is not the overlying point of the study. The study instead offers significant insights that increases in salaries lower incidents of theft and have a positive impact on employee honesty. Furthermore, other studies have shown that when salaries are cut, employee theft increases, shedding additional insights into this perspective. This is not to imply that higher wages automatically reduce theft, but discussing these studies may assist in advancing the knowledge base and help companies in developing the appropriate counter fraud strategies in this area.

Hotlines

Many companies establish and communicate fraud hotlines as part of their counter fraud program. This is seen as a very effective tactic to help thwart fraudulent incidents, both internally and externally. Many companies place great credibility in the use of hotlines to help identify and detect internal theft.

There is also merit in the use of hotlines in a strategy focusing on the external fraudster. Many insurance companies have established fraud hotlines and regularly publicize these phone numbers in various documents sent to policyholders and claimants. Hotlines can be effective, as they promote anonymous reporting and can help fraud departments develop leads into specific investigations. Although my personal experience with calls from hotlines is mixed, I would estimate that about half had merit and the other half were erroneous. Several specific hotline reports come to mind to exemplify this point.

One hotline tip came from a woman who claimed that her neighbors stolen vehicle report was fraudulent; she had seen the *stolen* vehicle after the neighbor reported the theft to the police. This tip resulted in the fraud unit opening an inquiry, and then an investigation. The tip was correct; the policyholder had arranged for a friend to steal the car, and then she filed a fraudulent stolen car report. After a successful investigation that proved the claim was fraudulent, the company denied the $35,000 claim, a very successful result. One of the most important aspects of these hotline tips is to assess the credibility and motivation of the caller, if possible. The hotlines should be promoted as anonymous, but there should also be the option for the caller to provide personal information if he or she desires. If the caller does provide callback information, the fraud unit should look to determine (and outright ask) what the motivation to report the fraud is. Oftentimes, the caller may have his/her own agenda, which causes wasted fraud resources as a result of the misleading hotline report. In the case mentioned above, it was later determined that the caller was motivated to contact the hotline, as she stated the policyholder was bragging about how she was going to get away with the fraudulent car theft and further boasted about how she was going to spend her claim check.

A second hotline tip had the opposite result; the caller claimed that a commercial building fire was intentionally set by the owner. This tip resulted in the fraud unit opening an investigation into the incident. After investigating, it was determined that the building fire was legitimate and that there was no fraudulent activity in the incident. Further inquiry focused on the hotline caller. With the help of the policyholder, it was determined that the caller was the owner of a competing business, who had the desire to taint the insured's reputation in the community. In this case, the hotline report resulted in many hours of wasted internal fraud resources, resources that could have been refocused to other areas. Hotlines can be a segment of a multipronged counter fraud approach, as they do have merit. However, carriers should be guarded with these tips, as some may be fictitious. The most important aspect in addressing the merit of these tips is to assess the credibility and motivation of the caller.

Information Sharing

Sharing information between carriers is seen as an area of fraud prevention that has the most opportunity for future development. My dissertation survey sample and many consultations with companies have revealed that there is little to no information exchange between carriers, and this creates many opportunities for future development. One of the basic foundations of academic research is information; we need information obtained from various sources in order to advance the knowledge base and help to create a strong infrastructure for solid decisions and strategies. Academic studies that are based on poor information are considered suspect, and applying their findings would cause serious validity and reliability concerns. One of the main challenges in information exchange in the fraud environment is the proprietary nature of the information being shared. When we consider that insurance carriers are a business and have profitability as their main priority, and not fraud detection, we have to understand that many companies will refrain from releasing information to prevent others from gaining a competitive advantage.

This perspective regarding profitability is extremely important when considering counter fraud strategies. In the current, highly competitive insurance environment, carriers are operating under tight combined ratios and are highly concerned with issues such as customer retention and loyalty. Thus, releasing any company-specific data is not encouraged and is downright forbidden in some carriers. We are therefore in a very interesting conundrum; we know fraud detection and investigation has a positive impact on profitability, as fraud units consistently show very impressive ROI results. However, fraud units need information to operate even more effectively, which the company is unlikely to release.

Lack of information exchange is a significant issue that bears attention. We understand, after post-9/11 analysis, that one of the significant flaws in our counterterrorism efforts in this tragic event was the lack of information sharing between agencies. Analysis of the response to Hurricane Katrina also illustrates the damaging effects of the lack of information exchange. As a result of many of these incidents, fusion centers were developed from this increasing need for information sharing between law enforcement and public and private safety sectors. Prior to these fusion centers, the three sectors (law enforcement, private, and public) acted and reacted to public safety incidents under separate and distinct management structures. It was determined that the three sectors could improve their global delivery of public safety through a fusion of information. In today's law enforcement community, information exchange is seen as a critical priority; one such example would be FBI's InfraGard, in which I am a participant. FBI InfraGard is a "partnership between the FBI and the private sector. It is an association of persons who represent businesses, academic institutions, state and local law enforcement agencies, and other participants dedicated to sharing information and intelligence to prevent hostile acts against the U.S."

We understand that information exchange in the insurance fraud industry is a priority, but given the proprietary nature of the information being released, we also understand the challenges therein. So what can companies do to help develop strategies in this area? There is an increasing movement toward more cooperation between companies in this area as they realize that the benefits of exchanging information outweigh the concerns. Membership in groups such as the National Insurance Crime Bureau (NICB) can facilitate the exchange of information in a more private, formal, environment. As discussed earlier, attending conferences, networking events, and industry association meetings can also facilitate information exchange on more of an informal basis.

Connecting with state agencies, such as the North Carolina Insurance Crime Exchange, can also assist with information exchange. European countries are also becoming increasingly aware of the benefits of sharing information, as I have witnessed certain areas that are exploring formal and informal exchange programs.

Exchanging information can be a very sensitive issue in our world of insurance fraud, but one that bears attention and consideration in order for us to move forward and advance our preventative efforts. We are not sure of the exact solution, but it is an issue that warrants further exploration and collaboration among companies.

Legislation

As part of a well-rounded counter fraud policy, it is very important that companies support legislative efforts in the areas in which they operate. Several larger carriers I have worked with have full-time employees that serve only this purpose: to monitor, develop strategies, and create a corporate legislative presence. Small to midsize companies may not have the resources to allocate full-time efforts; however, as long as some presence is realized, and some strategy developed, they will see benefits result. There are also several nonprofit agencies, such as the Coalition Against Insurance Fraud, that can assist with legislative strategies. It seems that many companies overlook the value of supporting legislation, as it may seem as if the effort is far from an actual quantifiable result. I have heard executives and directors remark that legislation efforts do not create any measurable result and can therefore be difficult to support. Even though this may be true, I would strongly encourage that these legislative efforts become a focus for companies, as the effort does result in impact, at a much higher level.

The survey sample from my dissertation discussed legislation, and specifically how there needs to be a much more contrived approach to support these efforts. Participants expressed frustration that there is a lack of laws that would assist them in pursuing fraudsters, laws that would punish and deter criminality. Three participants discussed how there is a lack of lobbying efforts and how this has a detrimental effect on fraud legislation. The sample illustrated how Florida has passed aggressive legislation with the support of lobbying groups, and as one participant claims, "Florida's fraud rates are declining rapidly; they have good politics there; here [New York], they are not good." Two participants disclosed how current legislative support is weak, and pointed out that even when there are laws pending, they get watered down, bogged down to the point they are completely ineffective.

Several participants, one who is currently a fraud attorney, discussed how there is a huge inconsistency in the method that the courts handle fraud cases; he further disclosed how the court seems to change its approach daily, making it very difficult to try to predict what a ruling may be. Fraud cases that are tried are so inconsistent, the participants mentioned, that some judges will rule one way and then their opinion is the exact opposite in the next case. A former police officer, one member of the dissertation sample, felt that this lack of support could be rooted in the reporting requirements of police departments. He explained that insurance fraud is a FBI Part 2 crime, taking a backseat to the Part 1 crimes, which gain all of the glamour and attention.

Mandatory auto photo inspection is another area that needs our legislative support. Under these statutes, policyholders are required to have their automobiles inspected prior to the issuance of a policy. This deters insurers from insuring vehicles with prior damage or vehicles that do not exist, a significant fraud concern. Below you will find a chart listing the states that have mandatory photo inspection requirements as of the last quarter of 2016.

State	Mandatory Inspection
Alabama	
Alaska	
Arizona	
Arkansas	

PSYCHOLOGY OF FRAUD

California	
Colorado	
Connecticut	
Delaware	
DC	
Florida	**XX**
Georgia	
Hawaii	
Idaho	
Illinois	
Indiana	
Iowa	
Kansas	
Kentucky	
Louisiana	
Maine	
Maryland	
Massachusetts	**XX**
Michigan	
Minnesota	
Mississippi	
Missouri	
Montana	
Nebraska	
Nevada	
New Hampshire	
New Jersey	**XX**
New Mexico	
New York	**XX**
North Carolina	
North Dakota	
Ohio	
Oklahoma	
Oregon	
Pennsylvania	

Rhode Island	**XX**
South Carolina	
South Dakota	
Tennessee	
Texas	
Utah	
Vermont	
Virginia	
Washington	
West Virginia	
Wisconsin	
Wyoming	

(Coalition Against Insurance Fraud, n.d.)

In this chapter, we discussed tactics that agencies and carriers can use in the development of strong counter fraud efforts; developing a multipronged approach will serve to create more touch points with fraudsters that will undoubtedly reduce fraud occurrences. Consider implementing these strategies in full or in part in order to develop a strong global anti-fraud approach. These specific strategies are based on the criminological theories we have discussed throughout this book, which focus on reducing opportunity and making fraud more difficult. In the next chapter, we will introduce you to a self-developed approach to fraud fighting—the behavioral bridge—and discuss how we can merge data and strategies to create an extremely powerful counter fraud program.

CHAPTER **6**

The Behavioral Bridge

Behavioral Bridge: Defined

This book is designed to open a new perspective into counter fraud, one that merges the world of criminology and economic crime. In the previous chapters, *The Fraud Triangle, Counter Fraud Efforts,* and *Developing a Multipronged Approach to Counter Fraud* we explored specific strategies that can be considered as one develops vigorous preventative measures. It has been demonstrated that there is significant value in considering the psychology of the fraudster as we develop focused efforts. In this chapter, I am very excited to reveal a self-developed approach to this endeavor, an approach that I have termed the *behavioral bridge*. This chapter will serve to fully illuminate this approach and where it fits in counter fraud.

We have clearly demonstrated that human behavior can provide us with valuable information in fighting fraud. We have explored many criminological theories and learned how a deeper understanding of the human mind can help fine-tune the methods we may utilize in our respective companies. I will propose a process whereby specific data on human behavior can be extremely useful in the development of strong strategies. Furthermore, we will explore how this data can be inserted into a data analysis or software system to help us assess the risk level of an individual and see if he or she has a higher propensity for fraudulent activity. I have termed this strategy a *behavioral bridge*.

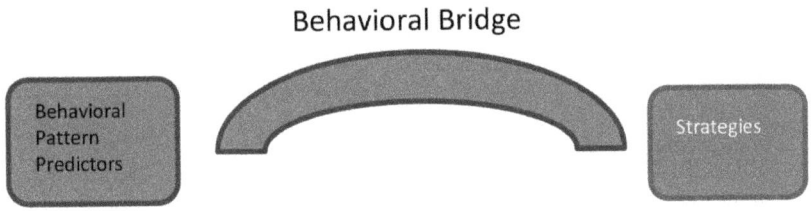

In its most simple form, the behavioral bridge is defined as the insertion of human behavior patterns into data or analytical systems in order to increase the effectiveness of strategies. Creating a behavioral bridge involves utilizing data from behavioral pattern predictors and inserting them into counter fraud strategies; the key element in this approach is the strategic utilization of this behavioral data. Before we discuss specific data applications in economic crime, let's explore how this concept is applied in other areas, such as counterterrorism.

There is no doubt that airport security remains a constant priority for law enforcement and security professionals, and human patterning is used frequently to assess the security risk of an individual, which also is referred to as profiling. Security experts are trained to look for patterns of behavior, or patterns outside of the norm (outliers) that may be indicative that an individual may be attempting to engage in questionable behavior. For good reason, many of these *flags* are not published due to security risks, but some examples could be an individual that is perspiring, someone wearing excessive clothing on a hot day, inconsistent eye shifting, fidgeting, excessive luggage, or light luggage.

Behavioral Pattern Predictors

The benefit of using behavioral predictors in practical application is exemplified by the identification of shoe-bomber Richard Reid at Charles de Gaulle Airport. Behavioral recognition revealed that Richard was not traveling with any luggage on a trans-Atlantic flight, which prompted security officials to question him prior to boarding. After officials confronted him, which did not reveal any other indicators, he was allowed to board the plane, where he attempted to light explosives from his footwear. Passengers intervened, which prevented an incident. This example illuminates the value of using predictors and how they can assist with strategic planning.

Facial recognition systems also have been used in various areas to add to the behavioral analysis field. We know from extensive studies conducted in this area that these predictors can be a strong indicator of potential threats, helping to prevent future terrorist activity. Using these patterning techniques is not without debate, as the American Civil Liberties Union (ACLU) has raised serious concern regarding these patterning techniques and how they may violate one's civil liberties. Now that we have established a framework of thinking for applying behavioral predictors, let us explore its application in a counter fraud setting.

Many sources of data are available to companies operating in financial services. It is important to consider how these internal or external data sources can be used singly, paired, or in groups, in order to provide extremely useful information to fraud departments. Consider all of the concepts discussed in this book, including activities aimed at identifying the highly vulnerable areas of opportunity within your company. Then creatively consider what data can be useful to assist with developing stronger approaches toward fraud prevention in these areas. Once data sources are identified that reveal behavioral pattern predictors, they must be bridged into application by merging them with fraud systems, preferably data or software systems. Let us explore some specific applications of the behavioral bridge.

Behavioral Bridge Applications

It also is important to consider all of the criminological concepts discussed in this book as one ponders this bridging approach, such as vulnerabilities, red flags, routine activities, rational choice, and the fraud triangle. As we have learned from the fraud triangle, strain theory, and many other theories, fraudsters commit crime because they experience pressure from different sources, such as financial pressure from job loss, debt, or growing family expenses. A behavioral pattern predictor (BPP) in this scenario would be any form of data that would be available to help identify individuals in this dire financial situation. In other words, any data source that would help to reveal an individual's financial behavior would be extremely useful if inserted into an anti-fraud solution that was based on a scoring system. Here are some examples of behavioral pattern predictors:

1. Credit score (FICO)
2. Bankruptcies
3. Liens and judgments
4. Default or late on loans—auto, tuition, boat, and mortgage
5. Default credit card bills
6. Default alimony payments
7. Late or default insurance premium payments
8. Late or default tuition payments

9. Recently unemployed
10. Recent health diagnosis resulting in expensive treatment

Now is where it gets fun! Consider all of the BPPs above and how they can be bridged into practical application. Remember—think creatively. Consider a company that has a software fraud solution based on scoring; that is, there is a defined set of rules that the company has identified as high risk indicators. Some examples of rules would be claim within ten days of new policy, late-night accident, accident with no police or witnesses, and so on. Every claim that is reported goes through the anti-fraud software solution and receives a score, with the highest scores gaining attention from the fraud department. These rule-based software systems are a highly effective counter fraud strategy, as they are working 24-7 at assessing the risk within a company.

In our sample fraud scenario, let us say that the company writes multiple lines of business—workers' comp, auto, homeowners, health, or property—and all of these claims are filtered through the anti-fraud solution. Traditionally, these software programs operate only within a specific line of business; that is, they only talk within the auto line of business and do not interact with property data. Now, ponder the value of creating a behavioral bridge between the data in these different lines of business. We have already established that unemployment can cause increased pressure, pressure that could act as a motivator to commit a fraudulent act. If a company has data (BPPs) in the workers' comp line of business, it will have information on certain individuals that are out of work for various reasons. What if we were to bridge the data from the workers' comp line of business (individuals out of work) with data from the auto line of business (individuals filing an auto theft claim) and create a higher risk score for these individuals? There is no doubt that this would create an incredibly valuable bridge that would result in more effective fraud identification.

If we continue with our auto theft example, we understand that financial motivators are a strong predictor. Ponder the value of bridging all auto theft claims to a behavioral pattern predictor data source, such as a database that tracks default loan payments or credit scoring.

Consider the limitless possibilities of bridging data into real-time applicability by inserting it into an anti-fraud solution. Behavioral patterns have been used by the Savings and Exchange Committee for years to monitor insider trading. The SEC looks for individuals who have a pattern of buying or selling stock in advance of a market fluctuation, such as an individual that repeatedly sells a specific corporate stock before it reduces in value. The SEC uses trading data as a BPP and bridges this into data systems for filtering and extremely accurate identification of questionable trading.

The Internet of Things (IoT) also will bring us increased opportunity for behavioral bridging. The IoT is on the radar of many companies as a potential future area of concern, with many organizations not entirely sure how they will be effected by the increased connectivity of the world. I recently completed a research article on the IoT and addressed the potential areas of interest for those operating in our industry. It is interesting how the availability of IoT data could assist companies if this data is bridged into applicable software systems. For example, what if the behavior of policyholders was tracked via wearable computing (Apple Watch, etc.) and this information was bridged into the software system of that individual's health insurance carrier. This anti-fraud software system could score an individual for premium assessment by risk level based on health condition, as tracked by the wearable computer.

Consider the usefulness of an automobile that is connected and how the policyholder's driving patterns (BPP) can be used to accurately assess premiums and also help in the investigation of a specific claim by revealing speed and direction of travel. Or consider social media, or how about using data gathered from Facebook, Twitter, and Instagram and inserting it into a data filter for interpretation.

I recently worked with a company in Europe where we explored the use of Google maps and the latitude and longitude coordinates to map the initial loss location. This initial loss location was cross-referenced with the closest hospital, which is where treatment would logically be received by any occupant in the accident. We then looked for any occupant that received hospital treatment farther than the close hospital, which could be an indicator of potential fraud. There are also additional opportunities to use this geographic information system (GIS) data. Mapping a patient's home address and cross-referencing with treatment facilities could be of use to fraud investigators. This could reveal potential suspicious treatment if a patient is passing a dozen medical facilities to receive treatment. This is an example of applying behavioral factors into a data system (i.e., the behavioral bridge). Again, be creative, as the possibilities are limitless.

In auto theft cases, it is a common investigative protocol to ask the insured for his/her cell phone records in order to assess the volume and frequency of calls before and after the actual time of theft. What if this information was cross-referenced with that individual's social media postings, where we could determine his or her exact geo location in relation to the auto theft. This could reveal strong indicators of that individual's involvement, or noninvolvement.

Admiral Insurance in Britain looked to use complex algorithms and Facebook to help it price car insurance. The hypothesis was that those individuals that post short, accurate sentences, such as "let's meet Monday at five thirty at Joes," as opposed to vague phrases such as "let's meet tonight" on Facebook would be more indicative of careful driving, as they were more precise. This seems like an incredibly creative behavioral bridge to use social media scoring as part of a pricing strategy. However, further investigation revealed that Facebook's policy prevents using data to determine eligibility for products, therefore putting an end to this creative pricing approach.

In Chapter 5, we discussed the development of a multipronged approach to fraud prevention. Consider any of the topics presented in that section as a potential behavioral bridge. We discussed employee vacation time and how encouraging time off can help filter internal fraud schemes. The research supports the contention that internal scammers are less likely to take vacation time, as they feel uncomfortable that others will discover their misdeeds in their absence. Unused vacation time can be a BPP that could be inserted into a data system for filtering, and identifying those internal employees that have a lower than average use of vacation days.

In my consultation activities, I also overlap in areas of credit card fraud, anti-money-laundering (AML), and counterterrorism financing (CTF) programs. These industries readily use big data and BPPs as part of their counter fraud strategies. For example, what if a credit card company were to track social media postings, or airline bills, and come to know exactly when a customer is on a plane or train. This data could then be cross-referenced with charges; showing fraudulent charges that occurred when the individual was unavailable and was actually on the plane.

As a final thought, the new age of fraud fighting is uncharted water, so a company must ensure that it complies with all ethical and legal rules and regulations when harnessing this data; this is where expert consultation is a must. I cannot stress enough the trend in counter fraud strategies toward the need to adapt to the use of data in fraud programs. In the dozen conferences (both domestic and international) I spoke at during this past year, the main theme in almost all of these events was that companies need to learn how to use data in order to survive the *new age* of fraud fighting. There are tremendous opportunities to use data, and we as fraud fighters must be able to harness and leverage this tool in order to increase our effectiveness at countering fraud.

As stated, behavioral bridging has almost infinite possibilities. As big data continues to grow and the Internet of Things continues to increase connectivity at an alarming rate, the opportunity to bridge these behavioral pattern predictors to strategies also will continue to advance. The critical perspective to consider in this bridging approach is to apply the behavioral aspects of individuals and then insert these behavior patterns into analytical systems for filtering, risk assessment, analysis, and preventative efforts. There is no doubt that creating these bridges will result in highly effective strategies in this world of big data.

CHAPTER 7

Final Thoughts

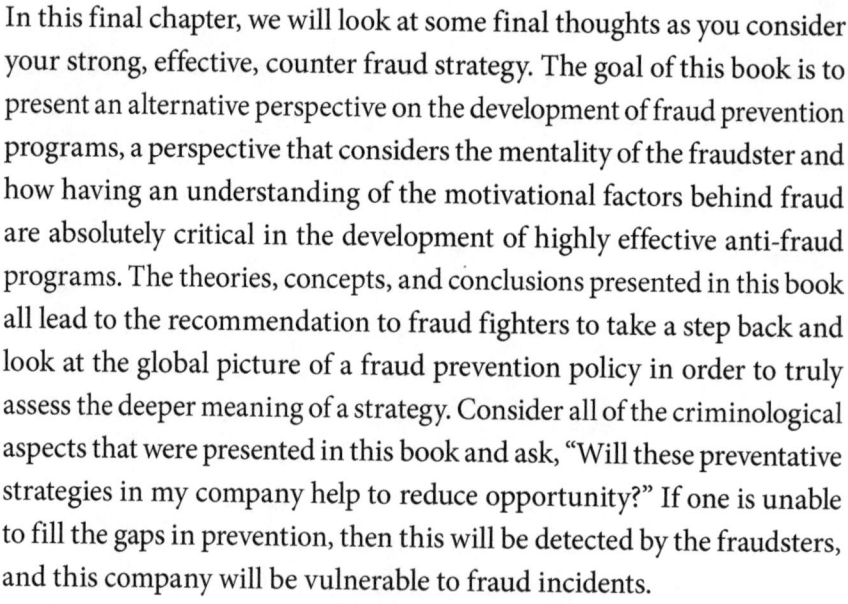

In this final chapter, we will look at some final thoughts as you consider your strong, effective, counter fraud strategy. The goal of this book is to present an alternative perspective on the development of fraud prevention programs, a perspective that considers the mentality of the fraudster and how having an understanding of the motivational factors behind fraud are absolutely critical in the development of highly effective anti-fraud programs. The theories, concepts, and conclusions presented in this book all lead to the recommendation to fraud fighters to take a step back and look at the global picture of a fraud prevention policy in order to truly assess the deeper meaning of a strategy. Consider all of the criminological aspects that were presented in this book and ask, "Will these preventative strategies in my company help to reduce opportunity?" If one is unable to fill the gaps in prevention, then this will be detected by the fraudsters, and this company will be vulnerable to fraud incidents.

In this final section, we will focus on some final thoughts as you continue your journey toward powerful counter fraud strategies.

Fraud Is a Broader Social Problem

In this book, we started with a theoretical approach toward fraud prevention and then transitioned into discussing some very specific strategies for application. Within our respective companies and agencies, we can take appropriate action and implement anti-fraud policies and procedures to help reduce fraud occurrences. As recommended, any preventative action is better than no action, as fraudsters will sense this vulnerability and act accordingly. However, there is growing consensus among those operating within the counter fraud community that fraud is a larger global problem that requires much more attention than specific company policies and procedures. One of the main themes from my dissertation sample was that fraud is a larger social problem, one that requires an adjustment of highly engrained thoughts that fraud is acceptable and will be tolerated by the general public. Many felt that specific preventative efforts only serve as a temporary, localized fix to the issue and that for fraud to truly be tackled, one has to fix the social problem of its acceptability.

There have been parallel comparisons to fraud and drunk driving from a social perspective. Drunk driving laws and procedures did not significantly change until there was a larger social movement, directed by Mothers Against Drunk Driving (MADD) in the 1980s. Until that time, driving drunk caused many deaths, but it was still not taken seriously and was not considered socially unacceptable. Once MADD began its high degree of involvement in political action, social awareness, training, and showing the true impact of the deviant action, then it became socially unacceptable, and that was engrained in our culture. It wasn't until that time that serious strides in prevention gained significant ground. Drunk driving is still a problem, but from a social perspective, there is an understanding that it is unacceptable behavior.

There is no doubt that fraud has impact, and those of us operating in the industry are well aware of the damaging effects; it should be our responsibility to communicate this damage to the public as much as possible in an attempt to slowly change the social acceptability of the crime. As we have demonstrated, many factors contribute to the social acceptability of the crime, including rationalization and entitlement theories. What is worrisome from my perspective as a professor for over ten years is that I have seen a disturbing trend in the younger generation; they generally feel more entitled than in years past. We are not sure where this will trend, but we are sure that this will definitely be a cause for concern as we attempt to reduce the social acceptability of the crime.

One of the main themes of my dissertation study on fraud prevention was: "Insurance companies are their worst enemy." The dissertation sample expanded on this and offered that insurance companies should be concerned with their appearance to the general public, specifically focusing on how the communication of high profits, high salaries, parties, and new buildings, has a detrimental effect on fraud prevention. I recall one life insurance company that built a state of the art facility close to my hometown. The local newspaper and media covered the construction of this building on almost a daily basis, showcasing the marble foyer, expensive artwork, and other lavish appointments. I overheard several neighbors comment negatively on this wasted expense, stating that if they had a policy with this particular insurance company, they would be quite upset at the high expenses. Companies should be very reluctant to showcase any form of expenditure other than those that support or benefit the policyholder.

Another example also serves to support this contention. A close friend is a medical doctor and is highly respected in his particular field of medical expertise. One colloquial discussion about the medical field transitioned into a debate on medical billing fraud and coding exaggeration. The doctor claimed that they (those in the medical field) have to consider the billing and diagnosis codes whenever they recommend treatment or submit a bill. He explained that if the diagnosis codes are not in line with the treatment, then the bill may be denied, costing the doctor's office time and money; thus, it is advantageous to have a diagnosis code that will serve to have the best chances of being approved by the insurance company's billing department. He then expressed extreme frustration over this practice, claiming that the health insurance companies are now driving treatment, not the doctors. In case you are wondering, our conversation became very lengthy and robust as I explained to him the dangers and negative impact of the fraudulent process he described. He rebutted and stated his actions were justified by referring to the fact that insurance companies are seen as rich and always trying to take advantage of them. Similar autonomous research has confirmed this alarming justification within the medical field. There are ways to slowly make fraud an unacceptable crime, and we have discussed some of the perspectives to this endeavor.

Application of Vulnerability Theory

It is the central theme of this book to promote the reduction of opportunity within a private or public sector framework in a manner that makes it more difficult for fraudsters to commit their deviant acts. We have discussed the mentality of the fraud offender and how these individuals look for any vulnerability within a counter fraud system to manipulate and capitalize upon. As we have discussed in the detailed vulnerability assessment section, there are methods a company can impose to avoid being seen as a soft target by fraudsters, thus making themselves less vulnerable to fraud occurrences. At this final stage of the book, we will discuss my theoretical approach toward fraud prevention, which can be summarized in what I term *vulnerability theory*.

The term *vulnerability theory* has been applied in many academic circles, such as emergency management, risk management, and government analysis. In the emergency management field, this theory seeks to explain why certain communities, companies, and individuals are more susceptible to disaster than others. In a government application, the theory postulates that vulnerability is a natural part of the human condition and that it is the responsibility of the government to ensure that all people have equal access to resources and programs, which must be distributed accordingly.

Vulnerability also has been studied in the area of fear and

victimization. In several compelling studies, the researchers focused on the cognitive perspective of an individual's fear of crime and victimization, attempting to draw conclusions along gender lines. These studies found that females are found to worry more about victimization because they felt less able to defend themselves and viewed the likelihood of victimization higher for themselves than those in other peer groups. Younger individuals also were found to worry more about victimization than older individuals.

The most applicable aspect of these vulnerability studies in a fraud setting does not pertain to the statistics and numerical results, but on the finding that the cognitive awareness of potential victimization was a key element to assist with victim avoidance. Furthermore, the simple acknowledgement of potential victimization was one key element for successful prevention in this scenario. Thus, if we draw upon this theory, simply acknowledging that we, as companies and agencies, may be vulnerable will assist with preventing fraudulent activity. As long as we create this awareness and recognize that we may be vulnerable, this is the first critical step in making us less of a target to the criminal element.

Vulnerability theory in an insurance fraud application is based on the tenants of both environmental theory and self-control theory. Environmental theory proposes that environmental factors influence behavior, and these factors can be controlled and manipulated. Self-control theory is based on the premise that those with low self-control, when placed in a situation of opportunity, will commit deviance because the target is vulnerable. Sociological theory focuses on the offender and how societal factors can cause fraudulent behavior. If we merge all of these principles together, we have an approach whereby the vulnerability of a target (insurance company or agency) promotes an environmental situation that an offender with low self-control will take advantage of; in other words, fraud will occur.

The decision to focus on this theoretical area is based on the author's professional experiences and direct observations, which provide support for its principles. Early in my career as a field investigator, I directly observed a linear relationship between auto theft and opportunity. As I would interview the victims of auto theft, it became apparent that some of them, in some manner, seemed to create an opportunity or increase the vulnerability of their vehicles, which ultimately led to their vehicles being stolen. Some examples of these opportunities that I observed were: leaving keys within the vehicle, leaving the vehicle running, parking in certain areas, easy access to keys within the homes, and valet parking. In effect, they seemed to create an environmental situation that when aligned with an individual with low self-control would entice that person to commit crime. The main tenet of this theory is to stress how environmental opportunity can be reduced, by focusing on ways that companies and agencies can decrease vulnerability in order to decrease their chances of victimization.

Vulnerability theory can be applied within the premise of several related criminological theories as discussed in this book. As we learned, routine activities theory is based on the tenet that a suitable target, lack of capable guardian, and motivated offender will lead to a criminal act. Vulnerability theory will focus on one element of this continuum: the suitable target. Cohen and Felson, in their seminal work in this topic area, posit that the best way to lower crime is to not focus on the offender, but focus on ways to reduce the opportunity to prevent crime.

Vulnerability theory will be assessed in several ways. As discussed earlier, the theory will focus on decreasing the opportunity for crime, which will reduce the number of instances that an offender with low self-control will have opportunity to commit crime. Drawing upon Hirschi's self-control theory, vulnerability theory will assume that all criminals are motivated, and the key factor in crime causation is the ability for self-control. Secondly, vulnerability theory has similar components of routine activities in that both will focus on the intersection between the offender and the target. Vulnerability theory postulates that an absence of a suitable target will reduce the opportunity to commit crime and therefore reduce the number of crime occurrences.

The Deterrent Effect

We have discussed and explored deterrence theory and its importance in an insurance fraud prevention setting earlier in this book: identifying the need for fraudsters to see that fraud will not be tolerated by companies, and clearly showcasing the risks of engaging in this deviant behavior. Deterring offenders from future fraud activity is viewed as one of the most significant goals for counter fraud professionals; research has revealed that these individuals make conscious choices to partake in criminal behavior. This choice-focused behavior is deeply rooted in concepts of cognitive theory, which posit that motivation and decision making are the main predictors of criminal behavior. Internal motivation and state of mind are the focal point of cognitive theory, which is similar to various foundational, traditional criminological theories, such as the classical and rational choice schools of thought.

The classical school proposes that crime occurs when the benefits outweigh the costs, when people pursue self-interest in the absence of effective punishment, and the decision to commit crime is a free-willed choice made by the individual. Cesare Beccaria and other classical theorists thought that all offenders are rational thinkers that make conscious choices to participate in deviant behavior. Building upon this cognitive focus of criminality, other theories, such as deterrence and rational choice, emerged. Both of these theories hypothesize that choice is influenced by costs and benefits, or its rationality; if costs are raised and more effort is needed to commit the crime, then crime will more likely be deterred.

Deterrence theory emerged as a significant crime theory in the 1970s when criminologists and economists, such as Gary Becker, suggested that deterrence should be a central focus of crime control strategies. Deterrence theory, building upon classical theory, focuses on the impact of deterrence as a punishment option, and specifically on how an individual will make a conscious choice to refrain from crime if the punishment is swift and severe (Lazear 2015). Deterrence theory suggests that specific and general deterrence reduces the likelihood of criminality and deviant behavior. Specific deterrence focused on the specific offender, and general deterrence focused on the general public. Cognitive theory, rational choice theory, deterrence theory, and the classical school of criminology all focus on the impact of the severity of crime as a control strategy. They assume that crime can be controlled by implementing programs that ensure swift and severe punishment and further argue that people will be deterred from crime if the pain associated with punishment outweighs the pleasure associated with the crime. In order to be truly effective, it must be swift, certain, and severe, and would favor clearly defined laws, strict judicial rules, and public displays of punishment.

What does the academic literature tell us in regard to the effectiveness of deterrence as a prevention strategy? The results are mixed, but are positive toward counter fraud efforts if placed in theoretical context. Capital punishment has created debate for years, with supporters claiming that the law of retaliation, or an eye for an eye, prevails, and punishment should be severe. Detractors claim that capital punishment is too severe, inhumane, and immoral. In-depth studies on this topic have failed to prove that implementing the death penalty actually deters crime, thus causing more heated debates. Other studies I have reviewed also show weak support for deterrence as a solid preventative approach. However, it is important to consider the criminological perspective of rational choice and free will to understand the effectiveness of deterrence in our fraud industry.

We have established that most fraudsters operate as rational thinkers and make conscious choices to engage in or refrain from deviant behavior. Operating on this assumption, the development of programs and policies that are severe should in fact have a deterrent effect. Therefore, it is likely that focusing on deterrence in a counter fraud setting will in fact have a positive impact on deterring fraud. Crime is seen as developing from two main factors: the pursuit of one's own self-interest and the lack of adequate punishment as a deterrent; thus, if we outweigh self-control with severe punishment, fraud should be prevented.

Measurement of Fraud

As an industry, there is consensus among us that the inconsistent measurement of fraud is one of the greatest barriers facing us in today's environment, as we are not sure where to focus efforts and are not clear on the true impact of the problem. In addition, there could be a significant ripple effect of our fraud values in stock prices, the allocation of resources, and internal and external staffing; thus it is a very worthy time investment for fraud fighters to focus efforts on fraud measurement. I have personally been involved in deep discussions on fraud measurement during training I was performing, on consultations, in focus groups, at conferences, while performing research, and even in colloquial one-on-one conversations, where is it evident that there is extreme confusion and disparity on the method we each utilize to quantify fraud.

To exemplify inconsistency, I recall one consultation meeting with a large global carrier that illustrates this point. The meeting attendees were all heads of fraud from different geographical areas, and the topic was the measurement of fraud. Even though the carrier had universal definitions of fraud and measurement from its headquarters, each manager interpreted these differently, which resulted in contrasting measurements. This caused a great deal of confusion (and arguments), as a fraud comparison between areas could not be realized due to the different perspectives.

I have been exposed to many different companies in many different regions and can personally attest to the perspective that there is no universal or common measurement of fraud. Considering that there are different-sized companies that operate in different areas of the country, or world, which write different lines of business and have different risks, it would be impossible to create one universal measurement that will work in every company.

However, even provided with these challenges, we are going to tackle this important issue. The purpose of this section is twofold: first, to start on a broader basis and introduce fraud measurement parameters based on three elements—the line of business, foundation, and formula—and, second, to then introduce a specific fraud measurement formula that is based on my experiences that could be considered for universal adaption. One of my highly respected colleagues sums up fraud measurement in terms we can all relate to: Mom's spaghetti! He explains that metrics and measurements of fraud outputs are like spaghetti sauce; until someone creates the perfect one, everyone's mother (meaning your program) will always make the best because it reminds you of home and reflects what is important to you. Thus, quite possibly we have all accepted that we will differ in terms of measurement, as we all have different priorities and strategies. However, this should not prevent us from having some sort of consistent parameters to use as a foundation, which will be discussed in the upcoming sections.

The first step in developing a more usable formula for fraud measurement is to reach an agreement as an industry that one specific, well-defined method may not work due to the aforementioned factors that make us all unique companies. How can we compare a large global carrier that writes four different lines of business and has an enormous counter fraud budget to a small carrier that operates in one state, has one line of business, and two fraud investigators? Thus, as an initial starting point, it is my recommendation that we first present a list of commonly used terms and then create definitions that we can all use as part of the measurement process. We can then delve deeper and develop certain universal parameters based on these definitions, which will ensure that we are all speaking the same language. This approach is used as a foundational element of sound academic research, the identification and definition of common terms. Even though our companies are all unique, using this approach will help to develop a consistent methodology toward measurement.

Most companies have some sort of measurement in place. However, I have consulted with several carriers where they had no tool at all. Again, some system is better than none, even if it is manual reporting! I am also aware of several companies that refrain from fraud measurement, as they fear civil repercussions if these metrics are publicized. In addition, for the companies that do have metrics in place, they create benchmarks on an annual basis, which is far too infrequent. Many situations could arise in those twelve months that could require attention and refocus. It is recommended that fraud units monitor their metrics at least twice a year, but more preferably once per quarter. This will provide adequate monitoring in order to react to new trends and patterns in the data.

It is recommended that counter fraud experts subscribe to the following three-step process when communicating fraud measurement so as to create a more consistent approach. First, identify the line of business; second, the foundation of measurement; and third, the formula of measurement.

Line of Business

When we speak of fraud measurement, it is recommended firstly that we identify and measure per specific line of business. I see too many companies that offer metrics that are too broad and do not offer the opportunity hone in on specific areas. A company can merge all of these LOBs together into larger percentages later if desired, but it is recommended that they first be calculated per specific line of business separately. Calculating them independently in this manner will provide a more robust and clearer picture of fraud within a respective company, which will undoubtedly assist with stronger counter fraud strategies.

When a carrier merges all lines of business together into one single fraud measurement, it will illustrate results that are simply not specific enough to make effective decisions. Specific lines of business could be: auto (motor), marine, motorcycle, RV, commercial property, commercial liability, product liability, personal property, renters, travel, dental, pet, medical/health, no-fault/med-pay, bodily injury, workers' comp, life, and so on. Measuring in this more specific manner by LOB first will make it easier for comparison across different companies and create more useful benchmarks for internal trending and analysis.

Foundation

After we identify the line of business, the second critical step is to establish the actual foundation of measurement to ensure further consistency. Carriers seem to differ in the actual item they are measuring, the most common being:

1. Claims. Some carriers use the number of claims as the actual foundation of their metrics, and do not consider that some claims could have one claimant and some claims have ten claimants; regardless, they still only take the file as one claim referral.
2. Features. Also referred to as the number of exposures. As mentioned above, if a claim has one claimant, this approach would count as one metric for measurement, but if a claim has ten claimants, this approach would recommend ten metrics.
3. Monetary value. Many carriers use the actual monetary value, or reserve, of a file as a foundation for their metric. Some use a statistical reserve, and others may use a manually entered reserve.

The importance of spending time reflecting and analyzing one's *foundation* cannot be stressed enough. It is critical, in order to illustrate accurate, strong final fraud figures that the foundation is accurately in line with corporate/agency strategy on a company and department level; the following illustrates this point.

I worked with one company that used *claim* as a foundation and helped the managers to understand that this metric did not showcase their true workload. This particular company wrote high-risk auto policies in metropolitan areas and where challenged with large-value medical provider cases with an unusually high number of passengers. A change to their metric foundation from *claim* to *features* helped them to, first, gain a clearer perspective on their fraud challenges, and second, also helped to more accurately showcase the hard work they were doing. For example, ten cases came in for investigation in one particular week;

five were assigned to John and five to Jane. So, using the *claim* metric, this would be: John-five, Jane-five for a total of ten. However, if we delve into the files further and use a *features* metric, it then looks like this: John-seven, Jane twenty-five, for a total of thirty-two. Investigation into these files revealed that several had a significant number of passengers. From a workflow perspective, using *features* was extremely significant, as we can now divide our investigations accordingly, and also from an impact perspective to others, the thirty-two is much more reflective and impactful than ten! As you can see by this seemingly simple change in foundation, the results where immense.

As an alternative, I am familiar with several companies that segment their files according to the type of claim, either single-claim investigation or major-case investigation. This is also an approach that will serve to consider the sensitive and unique challenge these different types of investigations bring. A major-case investigation often involves many entities and claimants and is difficult to quantify on a per claim basis, thus labeling them differently in the metric is a very effective approach.

Monetary value is also a very worthy metric; however, companies are encouraged to reflect internally on the validity and application of this approach. If a company is using a statistical figure to quantify fraud, where did this statistic originate and who developed the formula and monitors it? I have seen too often that this formula originates in numbers-oriented departments such as finance and actuary. These departments are extremely valuable and well-versed in metrics; however, they may not have an understanding of the unique environment, and challenges, of fraud occurrences. It also is important to ensure that these statistical formulas are updated and reflect current market conditions, as these change constantly.

If a company uses reserving as a metric, then the logical question is: Who comes up with the reserving parameters? I recall one company I worked with that used the initial reserve that the claims department listed on the initial report as its fraud metric. This reserve was extremely underrepresentative of the actual exposure, as many aspects of the claim had changed on the journey from initial report, to the fraud unit, to closure.

Another interesting and debatable topic arises when we speak of monetary savings, and that is one of future unforeseen fraud savings that is often a portion of a measurement metric. For example, in medical provider or health claims, we can accurately quantify savings by calculating the denials from submitted bills as a result of the investigation. However, some companies also will add a value into this savings based on the future unforeseen billed amount that the health provider will not be submitting as a result of this denial. It is very mysterious how this unforeseen figure is calculated, as there seems to be extremely different approaches to this measurement.

It is very important for companies to analyze the *foundation* that they are using as their measurement structure. This *foundation* must be aligned with corporate/department strategy, vision, and challenges. Fraud departments should trace the origin of these measurements and assess their validity, reliability, and practical application within the fraud environment in which they are each operating. Critically analyzing and adjusting one's *foundation* will assist with more precise fraud programs and a more accurate reflection to other units, departments, companies, and executives, the true picture of the hard work you are doing!

Formula

Now that we have identified two areas where we can be consistent in our fraud measurement communication (line of business and foundation), let us delve into the actual measurement formula to further define. To clarify, thus far we have established that the line of business and then the foundation be identified as part of a universal approach. So a conversation between companies at this point would consist of communicating the line of business and foundation—"motor line of business using a monetary value foundation," or "property line of business using features foundation." Now we move to the formula component, the last and slightly more conceptual segment.

The actual formula for fraud measurement varies significantly across companies and agencies, so much so that fruitful discussions concerning strategies are almost meaningless. Certain companies use outsourced software tools to gather information; others use internally built systems, and others use simple manual reporting. Each company has its own strategy, vision, budget, and fraud challenges, so it is expected that each will have its own metric for reporting and measurement. However, I have seen, more often than not, that most subscribe to their metric because it is their mother's spaghetti and something they are comfortable with and not because it is an effective measurement. I would encourage all carriers to spend time truly analyzing what their metrics are, and decide if they simply make sense.

We will now delve into actual measurements. Below is a compiled list of common formulas for fraud measurement based on my extensive exposure to various companies and agencies globally and also my review of appropriate industry and academic research.

1. Red Flag Rate. This is the measurement of the actual claims or exposures that breached the red flag threshold. It can be illustrated as a numerical value; if four thousand claims breached the red flag threshold this year, then the communication would be a four thousand red flag rate. This can also be shown as a percentage of the whole; if four thousand out of eight thousand total claims breached red flags, it would be communicated as a 50 percent red flag rate. The red flag rate is obviously dependent on the specific red flags; as such, the specific flags can be modified, which will significantly affect the red flag rate.
2. SIU Referrals, also known as Referral Rate. This is a very common measurement that shows the number of claims, exposures, or monetary value of the suspicious claims coming into the fraud unit. It is also recommended that this be further divided into the source of referral, such as from the claims

department, from a software tool, or from the internal fraud unit. This can also be expressed in multiple ways, but it is recommended that it is calculated as a percentage of referrals compared to the entire claims volume. If ten claims are referred out of a total claim volume of one hundred for the year, then this would be communicated as a 10 percent referral rate.

3. Acceptance Rate. Of the number of claims that were referred to the fraud unit (as above) this is the number, or percentage, that was actually taken and accepted for investigation. As can be expected, this value depends significantly on corporate philosophy. I recall one conversation with a fraud executive in a global location that mentioned that his acceptance rate changes monthly depending on his philosophy, with some months lower due to strained resources, and some months higher due to increased resources.

4. Conversion Rate. This is a commonly used formula that measures the number of cases successfully converted into proven fraud. This can be represented as a percentage or as a numerical value that is based on the number of cases accepted. For example, if one hundred cases were accepted for investigation, and five of those were denied due to fraud, the conversion rate would be 5 percent. There is a significant amount of disparity among carriers on what defines the word *convert*. Some use the parameters of claims where the fraud was proven and the entire claim denied; others measure if the case was prosecuted, and others if the file was mitigated. Each of these approaches may make sense in different carriers; however, it is recommended to ensure consistency that whichever of the three approaches is used, that it is clearly identified in communications. For example, ABC Insurance had a 5 percent conversion denial rate, or 5 percent conversion mitigated rate, or lastly, a 5 percent conversion prosecution rate.

5. Fraud Identification Ratio: The number of fraud referrals divided by the number of denied claims as a result of fraud

investigation. Fraud referrals are taken from both manual and software-assisted solutions.
6. Average Days to Investigate. Self-explanatory measurement that considers the time frame from the time the claim entered SIU until the file was closed. This can be measured per file or per investigator.
7. ROI—Return on Investment. This figure will consider the dollar savings of an investigation and also include the relative hard and soft expenses, such as salary, benefits, vendor expense, and so forth.
8. Yearly Operating Expense: Internal calculation based on the sum of all fraud-related expenses, such as salary, benefits, tuition, travel, training, company vehicles, cell phones, and so on.
9. Fraud Savings. Calculated using various methods, such as reserve of file, manual calculation, statistical reserve, statistical amount, and presented per claim, LOB, or business unit.

We have now established a baseline of formulas that are common in our industry and will serve to more consistently approach this important topic. Now that we have completed our three-step process, and are on our way to a consistent approach, a communication between companies would consist of identifying all three segments: the line of business, foundation, and formula, such as property line of business, claims foundation, SIU referral rate from claims of 5 percent; or medical line of business, features foundation, conversion rate from denials of 10 percent.

This three-step process is designed to facilitate a more consistent approach to fraud measurement, one that will assist with the visualization of a clearer picture of this mounting problem. As each of our companies is unique along many different variables, it is not feasible to develop one formula of fraud measurement that will fit into all scenarios, as this will cause unreliable results. I have directly experienced this when attempting to integrate one company's measurement perspective into another company; most often it is not effective and simply does not work. Thus, it is the recommendation to follow the basic parameters and process as outlined in this section, which will help to ensure that we are all speaking the same counter fraud language.

Fraud Measurement Recommendation

Considering this fruitful discussion on metrics, it would be very remiss of me to not introduce a measurement recommendation for fraud measurement as a component of this book. This recommendation is designed to focus on one line of business and serve as a platform for discussion and potential application. It is believed that this formula will be the most accurate reflection of true fraud investigations and serve to create a consistent approach to measurement so that companies can start speaking apples to apples (or spaghetti to spaghetti).

The recommendation will be as follows:

- line of business-personal auto
- foundation-monetary value
- formula-fraud savings

My recommendation is therefore based on personal auto, monetary value, and fraud savings as my structure. Personal auto is the most common and widely familiar LOB, so this was chosen for ease of application. Monetary value and fraud savings were chosen because these are seen as the most accurate and true reflection of the performance of a fraud unit; let's expand on these last two categories.

For monetary value and fraud savings, it is recommended that experts take a manually entered savings figure at the conclusion of an investigation and refrain from including future, unforeseen savings. I am aware that the recommendation of a manual figure will cause some fraud professionals to lose sleep tonight, especially considering that the world of counter fraud is transitioning into one of workflow, optimization, and data. However, I strongly believe that the human component, something we have spoken about many times in this book, is an absolute critical component to include in this extremely important issue of fraud measurement. No one knows a particular claim more than the actual investigator that is assigned, and therefore no one can more accurately quantify the true savings of that claim. Every single claim that I handled as a field investigator was unique, and therefore every single one would have different exposures and values, figures that I was the most qualified to establish.

It is acknowledged that one of the weaknesses and criticisms of my recommendation is that it is based on a subjective figure, and as such, less reliable, credible, and accurate than an objective one. But, I argue that this subjective figure will be much more reliable, as it is based on true operational level input, directly from the experts that have more knowledge about a specific claim than any other system can quantify. If developing a true, accurate fraud savings metric is our goal, than this manual approach is undoubtedly the most reliable, credible, and accurate.

Even though this is a self-quantified figure, we can still utilize technology to streamline this process so it does not cause undue additional time constraints on very busy investigators. I have seen companies that have a field text box in their software solution that prompts a savings amount at the closure of a case. Other companies use internal tracking and savings submission worksheets to help streamline the process. There are many options to assist with this tracking structure, but at its core, companies should ensure that the operational level investigators are the driving force behind these figures. There are many advantages to the other approaches that use reserving, statistics, and algorithms to calculate savings, but in my opinion they are not a true reflection of the savings result.

It is also recommended that fraud units take the actual and true savings based on that particular date and time of closing and refrain from including and calculating future, unforeseen savings in this metric. I operate in many other economic crime prevention areas, including the financial fraud sector, where this could be considered creative accounting, and potentially unethical and unscrupulous. I fully acknowledge that we need to take credit where credit is due, such as when we deny a medical claim and realize that our denial will result in no future bills being paid. However, if creating a true fraud savings metric is the goal, then future savings is much too subjective to include.

Fraud measurement, and the relative inconsistency, is one of the largest challenges facing us in the counter fraud world. Each carrier and agency has its own strategy, vision, and challenges, and accordingly develops its own measurement metric. It is difficult to find one approach that will work, but hopefully this section provided multiple options. First, if we do not use the same exact formula across companies, we can still use the same terms (talk the same language) and use the basic parameters of measurement as outlined using the line of business, foundation, and formula approach. Second, a specific formula using a personal auto, monetary value, and fraud savings metric was introduced, which can be immediately integrated and used in a counter fraud program. We are now going to move our discussion more globally and delve into international counter fraud and learn about some of the unique approaches to fraud fighting worldwide.

International Perspective

As a consultant, I am fortunate to have the incredible opportunity to travel internationally and research, analyze, implement, and train companies and agencies on approaches to fraud prevention. In this section, the goal is to present some of the unique perspectives to counter fraud in various parts of the world that I have been exposed to, perspectives that will hopefully provide an alternative view toward your own effective strategies.

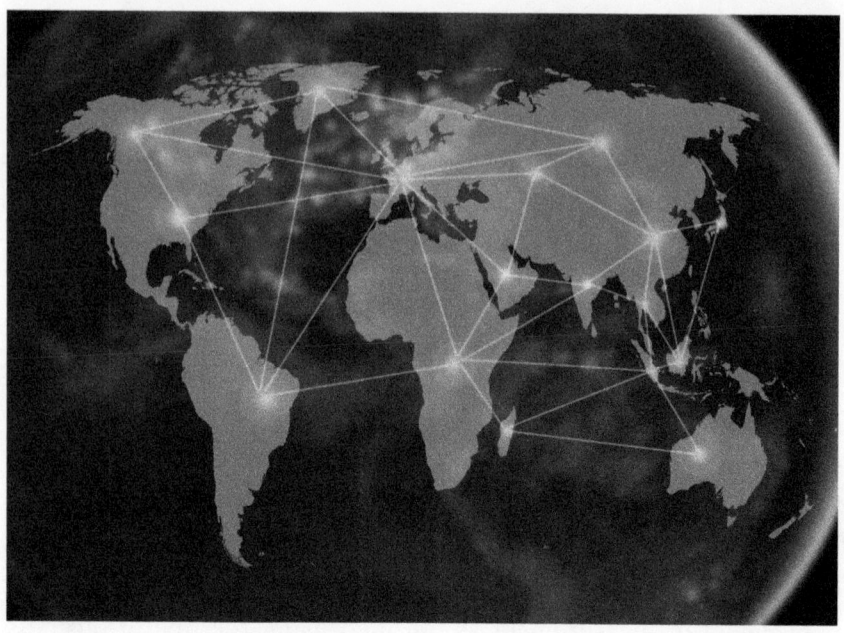

International Perspective

Statistics reveal that the majority of insurance premiums are collected in three areas of the world—31 percent in North America, 32 percent in Europe, and 30 percent in Asia—which indicates that fraud will be a major concern in these locations (Insurance Europe 2016). I have seen a global trend in companies and their increased focus on workflow models to optimize fraud and claim operations. This increasing focus on quicker settlements has resulted in surges in the area of opportunistic fraud, as these claims slip through the cracks of many counter fraud programs. This is where companies have seen great benefit with the use of software systems to assist with fraud identification and filtering of these fast-track claims. Certain software systems have the ability to effectively identify fraud and also optimize the claims and fraud process, making them extremely effective from many perspectives.

Looking at international trends on organized fraud, there is no doubt that organized criminals focus on those jurisdictions that have weaker and more vulnerable systems of controls, such as weaker regulatory standards, lower detection capabilities, and lack of information sharing. Analyzing global fraud patterns proves that these weaker areas gain more attention from fraudsters; companies that operate internationally must consider this as part of their global counter fraud strategy. There is also increased interest, or concern, about technology and how this will impact the future of fraud investigations. I have been asked by carriers from all areas of the globe how the Internet of Things will impact us; quite frankly, we do not know at this point, as this is still an area of exploration in our industry. Similar to social media sites such as Facebook, LinkedIn, and Twitter, and how these have changed our fraud investigations, the IoT is uncharted water and an area that we will continue to explore together as a counter fraud community. Let us now explore specific areas around the globe and delve into their counter fraud culture.

Canada

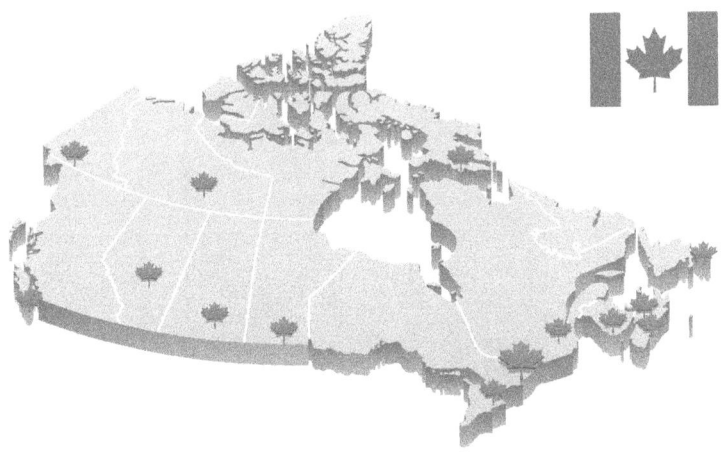

I have witnessed a rising trend in companies and agencies in Canada suffering increased fraud losses in the areas of underwriting and premium fraud. Several companies I have spoken with in the early part of 2016 have remarked that this area of fraud is their sole focus for the upcoming year, as they are suffering massive losses in this area. They indicate that rate evasion cases increased over 150 percent in one year, with policy residence misrepresentation at the top of the list. It appears that this spike in fraudulent activity is due in part to increasing premiums in metropolitan areas, making higher urban premiums a driving force as insureds seek lower-tiered areas in rural locations.

These carriers and agencies also report that approximately half of claims that were prosecuted for fraud also involved some aspect of underwriting fraud. There also appears to be an increasing trend in opportunistic fraud in Canada, with the Insurance Bureau of Canada (2016) reporting that fraud costs $1.6 billion per year—$1.2 billion in opportunistic fraud and $275 million in organized fraud. Fraud efforts should therefore focus on the opportunistic fraudster and methods to reduce opportunity as discussed in this book. Preventative efforts should target the higher volume, smaller value claims that are usually the primary characteristics of these opportunistic cases.

Malaysia

One of the most unique insurance systems I have been exposed to derives from an area of Southeast Asia—Malaysia. There is a distinct dichotomy of insurance in Malaysia; the two-part system is comprised of insurance and *takaful* policies. Both are extremely unique and differ from one another in the manner in which they are regulated, supervised, and integrated. The insurance products are managed by the Insurance Act of 1996 and follow the traditional transfer of risk between an individual and a company; policyholders pay a premium to carriers, who in turn reimburse the insureds if a loss occurs. In contrast, *takaful* is managed by the Takaful Act of 1984 and is governed by *Shariah*, or Islamic law. Under this program, policyholders will contribute a certain amount of money to a shared *takaful* fund (also known as participative contributions), and they then become members under contract. Under this contract, all customers agree to assist each other in the event another individual within the contract suffers a loss; these principles are based on the Holy Quran, which promotes assistance toward those in need.

It is apparent that these two systems cause increased challenges for those operating in a counter fraud setting in this area. Based on my interviews with several carriers and agencies in Malaysia, they report the following alarming statistics. Almost half of them reported they have experienced some sort of internal theft in the last year, with all of them remarking that this is an increasing area of concern. They further report that approximately 15 to 20 percent of their claims are fraudulent, significantly higher than the industry standard of 10 percent. They also indicate that opportunistic fraud is a large problem, with approximately three-quarters of their cases in this area and the other one-quarter in organized fraud. These trends in Malaysia will continue to cause demands on fraud fighters operating in this geographical area.

Italy

The fraud situation in Italy is unique in that there is a high propensity that an individual will be involved in an accident; one vehicle in twelve in Italy is involved in an auto accident as compared to one in twenty-three in France. There is also a higher percentage of injuries claimed in Italy; over one million people per year in Italy claim injuries as compared to under two hundred thousand in France. As a result of these startling statistics, premiums in Italy are extremely high, higher than many other neighboring countries, such as France, where premiums are approximately half of the cost. Premiums in Italy increased 18 percent between the years 2002 and 2009, with the average European country increasing only 7 percent. If we apply our knowledge of criminological theories to this environment of high premiums and high-risk drivers, we can hypothesize that an increased sense of justification and entitlement is probably felt by Italian policyholders, which results in increased fraud.

One of the most significant barriers to counter fraud efforts in Italy is the lack of accurate detection of fraudulent cases. Statistics from the Istituto per la Vigilanza sulle Assicurazioni—IVASS (2015) and other sources reveal that approximately 2 percent of claims are investigated for fraud in Italy, when fraud experts report that the fraud rate should be closer to 10 to 14 percent, indicating a huge gap in lost opportunity, and a gap that undoubtedly attracts fraudsters. Information sharing, or lack thereof, is also a major concern in Italy, with agencies just starting to assist with the facilitation of data exchanges. In addition, the regulatory environment causes increased challenges for fraud fighters, as there are stricter protocols for claim settlements and also tougher data privacy issues.

Germany

The German insurance association (GDV) estimates that fraud costs in excess of four billion euros per year, and approximately 10 to 15 percent of all claims are suspicious, making fraud a very significant issue in Germany (Mckinsey & Company 2015). For example, in the warranty line of business, research shows that almost half of all smartphone claims are suspicious and should be investigated.

We have discussed many times during the book how the lack of information exchange is a significant issue that should be a major priority of carriers, and Germany is above average in developing counterstrategies in this area. The German insurance industry information system (HIS) was developed as an information bureau to assist with fraud detection and to serve as a strong foundation for developing fraud intelligence throughout the country. The HIS database is regularly used as one component of a fraud prevention strategy, with users applying the information obtained as part of a fraud detection automated solution.

Workers' compensation in Germany differs from other countries, as there is an additional offering available to employees. The traditional workers' compensation system is modeled after Otto von Bismarck and is commonly known as the Bismarck System, which is funded entirely by employers. Under this system, compensation is provided to employees for work-related injuries in the form of medical coverage and continued benefits based on a percentage of medical impairment. There is also a second workers' compensation program in Germany, termed *Krankentagegeldversicherung*, or daily benefits allowance. Under this program, an employee is not entitled to medical bills, but only a work loss benefit, based on the degree of impairment and disability as prescribed by a medical doctor.

One of the most unique aspects of fighting fraud in Germany pertains to the presence of workplace representation in private-sector companies. Work councils have significant power and provide representation of employees within an organization; these councils are similar to unions but without the formal structure. The Works Constitution Act of 1952 (amended and approved again in 2001), established the legal basis for work councils in all private-sector organizations that have at least five employees. The chart below showcases the number of employees and the equivalent members of the work council required (Berlin Business Location Center 2016).

Employees	Number of Members of the Works Council
5–20	1
21–50	3
51–100	5

101–200	7
201–400	9
401–700	11
701–1,000	13
1,001–1,500	15
1,501–2,000	17
2,001–2,500	19
2,501–3,000	21
3,001–3,500	23
3,501–4,000	25
4,001–4,500	27
4,501–5,000	29
5,001–6,000	31
6,001–7,000	33
7,001–9,000	35
>9,000	add 2 for each additional 3,000 employees

The main goal of the work council is to ensure that the organization does not make decisions without considering the impact to its employees; basically, it is to strike a balance between employer and employee rights. Accordingly, all major decisions, including the implementation of a counter fraud strategy or program, must be presented to the council. In several implementations of technological fraud solutions, I have seen where delays, modifications, and even cancellations of these fraud programs were the result of work council review. This, therefore, becomes a major consideration for fraud fighters in Germany when they consider the development of a counter fraud policy.

I recall one specific consultation I was involved in whereby the executives in an insurance carrier were extremely concerned with the council's opinion of a software-based fraud detection system that was ready for installation. In this situation, the carrier was suffering significant losses due to fraud, as it did not have a fraud detection system to help identify suspicious cases and flag them for investigation. The carrier desired to increase its current fraud rate from 1–2 percent to 10 percent through the implementation of the program. The proposal was to install a powerful software program that would be able to identify and filter fraudulent cases. When the company's work council members reviewed the proposal, they had concerns, as they thought the software system might also be used to track employee performance such as claim settlement time, exposure amount, and so on. Modifications of the program ensued, and multiple detailed explanations also occurred, in order to ensure to the council that the main purpose of the program was for fraud identification and not employee monitoring.

German fraud fighters need to be aware of the interesting and unique relationship that is evident between employers and the council, and strive to maintain close communication with the council on the implementation of fraud detection policies.

Switzerland

Gross written premium in Switzerland totaled 54.4 billion euros in 2015, with claims and payments totaling 41.3 billion. Insurance premiums in Switzerland are among the highest in all of Europe; motor premiums averaged 664 euros, compared to the industry average of 225 euros.

Average property premiums in Europe in 2015 were approximately 150 euros; Switzerland was the highest of all countries at 457 euros. This trend continues in the health insurance area, where the average premium is approximately 200 euros, and the Swiss average is nearly 1,080 euros! Consistent with statistics in other areas, it is estimated that approximately 10 percent of all claims submitted in Switzerland contain some element of fraud, with motor, contents, valuables, and travel insurance being the top areas of concern.

From a legal perspective, insurers in Switzerland are given the authority to rescind a policy if a fraudulent claim is filed; this right is specifically granted in the Insurance Contract Act (VVG). The extremely high premiums in Switzerland result in a starkly different perspective on fraud. From a social perspective, I see a distinctive contrast in the fraud philosophy of Swiss insurers as compared to carriers in other jurisdictions. This distinction lies primarily in the insurers' opinion of their insureds; carriers consider their policyholders as highly valuable, loyal, honest members of their business contract. Furthermore, this creates an atmosphere where insurers may be hesitant to pursue, investigate, and accuse their insureds of fraudulent activity (Insurance Europe 2016; Swiss Insurance Association 2016).

This is not to imply that Swiss insurers are complacent in their approach to fraud; quite the contrary, they are aggressive in areas of fraud detection. However, their mind-set is one where they consider their insureds as part of their counter fraud policy. I have had very insightful conversations with many carriers that operate in Switzerland that have unanimously remarked that retaining their insureds is a high priority for their companies, and, furthermore, that this has an absolute impact on their counter fraud policy. One carrier specifically discussed how many of its insureds have been policyholders for many years, and how the company strives to develop lifelong relations with customers. This particular company measures its policy retention rate and publishes this. It is a very high priority internally within the various departments. There was no doubt in this company that customer loyalty was a much higher business priority than fraud identification and investigation. This particular company recently had uncovered several high-exposure fraud cases; however, a business decision was made to pay these claims and ignore the fraud, as these policyholders had been valuable customers for many years. As all of us know that operate in counter fraud, there are always competing forces between fraud and business, sometimes with the balance skewing toward the business goals of the organization.

Conclusion

We embarked on a journey in this book that took us into the exciting world of counter fraud. We first established that fraud is a very significant problem on many different levels. We understand that there is significant financial and humanitarian impact due to fraud, and therefore counter fraud efforts must be a high priority for those operating in this industry. Estimates reveal that the cost of fraud is approximately $80 billion in the United States, translating into $1,000 per year, per household. The financial impact to those companies operating in this industry is immense, with many organizations losing significant revenue due to the increased exposure in this area. Some companies report that this is the single most significant risk that they face in today's business environment, and they further remark that the problem is continuing to grow at a truly alarming rate. Also consider that many carriers are operating on a very tight combined ratio, and additional revenue loss will have significant impact on this ratio and decrease overall profitability. Couple this with an incredibly competitive insurance market, and we realize that organizations should make counter fraud efforts the most important priority in their overall business strategy.

Increasingly alarming is that many anti-fraud strategies are made without considering the psychological aspect of the economic criminal; this offender differs significantly from other criminals on many cognitive levels. We reviewed the different schools of criminology and saw how these are applied in many other criminal areas and how these concepts can also be applied and used strategically in a fraud setting. The sociological school of criminology is seen as the most applicable to us, as this perspective focuses on environmental aspects of behavior, and on how carriers can control certain external factors that will undoubtedly result in behavior modification. The most useful theories to counter fraud are: routine activities, rational choice, strain, and deterrence theories, as these are ideally applied in a fraud scenario. It was established that a rational criminal will make a conscious choice to partake in fraudulent activity; therefore companies should focus on reducing the opportunity to commit these crimes in their countermeasures. The risk-versus-rewards consideration is a constant theme in these sociological theories, and accordingly our focus should be on increasing the risk and reducing the reward, which will result in fewer instances of fraud.

A full summary of the fraud triangle was provided, where we explored the three main elements of this concept: pressure, motivation, and opportunity. We realized that the pressure to commit fraud can come from many different sources, but often lies in financial stress that pushes individuals toward fraudulent behavior. Motivation comes from many sources as well and is often explained simply by greed. The opportunity to commit fraud is the last element that was explored; we came to understand that if there is opportunity to commit an action, it will often ensue, as this is seen as a weak area within a company's system of controls. We then explored specific counter fraud efforts and integrated the criminological theories discussed. Identifying specific threats by performing a vulnerability assessment is seen as a critical step in the process; companies should take the vulnerability assessment very seriously, as it will clearly identify the areas that are at the most risk within an organization. This exercise is meant to reveal the most vulnerable areas within a company, areas that can be addressed accordingly. We know that we have limited resources in counter fraud, and the vulnerability assessment is a method for us to visually identify the top priorities so we may target these areas. It then is absolutely critical to take these areas of weakness and develop focused strategies. Many organizations seem to be content with the identification of vulnerabilities and desire to not proceed any further in order to translate these weaknesses into actionable policies.

Two of the main themes of this book are to reduce opportunity and develop a multipronged approach to fighting fraud. Developing single-stream programs will impact a single stream. However, we understand that fraudsters can be motivated by different factors and therefore can be thwarted by different approaches. We need to develop tactics that will hit as many points as possible and consider many different angles in our efforts.

A presentation of the self-designed behavioral bridge was discussed. This should be a perspective that companies consider in their approaches, bridging behavioral patterns into specific data and analytical systems in order to create a rich, robust data stream for application. Traditionally, behavior and data do not speak the same language. However, bridging behavior (and social information) is now possible through behavior pattern predictor systems that gather behavioral data and then offer it for insertion into a data system for interpretation. Behavioral pattern predictors can be any form of behavior that is measured in some capacity. The possibilities are limitless and growing on a daily basis; big data is rich with applicable information that should be considered in counter fraud.

We also understand that one of the most significant challenges facing counter fraud is the measurement of the problem. A lack of consistent measurement parameters is one area of vast improvement; thus a protocol for measurement was introduced in the final chapter. This approach toward measurement will focus on three areas for increased consistency: the line of business, the foundation, and the formula. Using this format will ensure that we are all speaking the same language and are calibrated in our terms and definitions.

The world of counter fraud is extremely exciting, and one that brings new challenges every single day. I have directly witnessed the similarities and differences in counter fraud approaches around the globe, intrigued by the methodology behind certain tactics. No matter the company, line of business, or geographical area, we should all be excited, as we will always have new fraudsters to fight and new approaches to consider, all keeping us super-sharp counter fraud experts. Have fun and stay safe!

References

Abramovsky, A. 2008. "An unholy alliance: Perceptions of influence in insurance fraud prosecutions and the need for real safeguards." *Journal of Criminal Law & Criminology* 98, no. 2: 363–427. Retrieved from http://www.law.northwestern.edu/jclc/.

Akers, R. 1991. "Self-Control as a General Theory of Crime." *Journal of Quantitative Criminology* 7, no. 2: 201–11. Retrieved from http://www.jstor.org/stable/23365747.

Alm, J., D. M. Bruner, and M. McKee. 2016. "Honesty or dishonesty of taxpayer communications in an enforcement regime." *Journal of Economic Psychology*, 5685–96. doi:10.1016/j.joep.2016.06.001.

Allgulander, C., and B. Nilsson. 2000. "Victims of criminal homicide in Sweden: A matched case-control study of health and social risk." *American Journal of Psychiatry* 157, no. 2: 244.

Andresen, M. 2006. "A spatial analysis of crime in Vancouver, British Columbia: A synthesis of social disorganization and routine activity theory." *Canadian Geographer* 50, no. 4: 487–502. doi:10.1111/j.1541-0064.2006.00159.x.

Antonaccio, O., W. R. Smith, and F. A. Gostjev. 2015. "Anomic Strain and External Constraints." *International Journal of Offender Therapy and Comparative Criminology* 59, no. 10: 1079–1103. doi:10.1177/0306624X14533071.

Ariely, D. 2013. *The Honest Truth about Dishonesty: How We Lie to Everyone—Especially Ourselves.* New York: Harper Collins.

Asmat, D. P., and S. Tennyson. 2014. "Does the Threat of Insurer Liability for 'Bad Faith' Affect Insurance Settlements?" *Journal of Risk and Insurance* 81, no. 1: 1–26. doi:10.1111/j.1539-6975.2012.01499.x.

Association of Certified Fraud Examiners. 2016. "Global Fraud Study." Retrieved from https://s3-us-west-2.amazonaws.com/acfepublic/2016-report-to-the-nations.pdf.

Association of Certified Fraud Examiners. 2016. "Sample Fraud Policy." Retrieved from http://www.acfe.com/uploadedFiles/ACFE_Website/Content/documents/Sample_Fraud_Policy.pdf.

Bandura, A. 1977. *Social Learning Theory.* Upper Saddle River, NJ: Prentice Hall.

Berlin Business Location Center. 2016. "German Works Council Constitution Act." Retrieved January 19, 2016, from http://www.businesslocationcenter.de/en/business-location/labor-market/employment-law-and-collective-contracts-system/german-works-council-constitution-act.

Brody, R. G., and F. S. Perri. 2016. "Fraud detection suicide: the dark side of white-collar crime." *Journal of Financial Crime* 23, no. 4: 786–97. doi:10.1108/JFC-09-2015-0043.

Cabinet Office Behavioural Insights Team. 2012. "Applying behavioural insights to reduce fraud, error and debt." Retrieved from https://www.gov.uk/government/uploads/system/uploads/attachment_data/file/60539/BIT_FraudErrorDebt_accessible.pdf.

Cable News Network. 2016. "Andrea Yates Fast Facts." Retrieved March 10, 2016, from http://edition.cnn.com/2013/03/25/us/andrea-yates-fast-facts/.

Carrabine, E. 2016. "Changing Fortunes: Criminology and the Sociological Condition." *Sociology* 50, no. 5: 847–62. doi:10.1177/0038038516645751.

Carter, S., S. Carter, and A. Dannenberg. 2003. "Zoning out crime and improving community health in Sarasota, Florida: Crime prevention through environmental design." *American Journal of Public Health* 93, no. 9: 1442–45. doi:10.2105/AJPH.93.9.1442.

Cartwright, A., and J. Roach. 2016. "Fraudulently Claiming Following a Road Traffic Accident: A Pilot Study of UK Residents' Attitudes." *Psychiatry, Psychology and Law* 23, no. 3: 446–461. doi:10.1080/13218719.2015.1080148.

Chamberlain, Mitchell. 2016. "Are Genetics Responsible for Criminal Behavior? Many Prisoners Share a Gene Linked to Personality Disorder." Retrieved October 1, 2016, from http://www.medicaldaily.com/are-genetics-responsible-criminal-behavior-many-prisoners-share-gene-linked-397741.

Charatan, F. 2002. "FBI investigates cardiac surgeries." *BMJ: British Medical Journal (International Edition)* 325, no. 7373: 1130.

Chen, C. X., and T. Sandino. 2012. "Can Wages Buy Honesty? The Relationship Between Relative Wages and Employee Theft." *Journal of Accounting Research* 50, no. 4: 967–1000. doi:10.1111/j.1475-679X.2012.00456.x.

Coalition Against Insurance Fraud. n.d. "By the numbers: fraud statistics." Retrieved December 1, 2016, from http://www.insurancefraud.org/statistics.htm.

Coalition Against Insurance Fraud. 2003. "Study on SIU Performance and Measurement." Retrieved from http://www.insurancefraud.org/downloads/siu_study.pdf.

Coalition Against Insurance Fraud. n.d. "Insurer Requirements." Retrieved July 15, 2016, from http://www.insurancefraud.org/requirements-grid.htm.

Coalition Against Insurance Fraud. n.d. "Four Faces Study." Retrieved March, 2, 2016, from http://www.insurancefraud.org/requirements-grid.htm.
http://www.insurancefraud.org/four-faces-study.htm.

Cohen, L. E., and M. Felson. 1979. "Social Change and Crime Rate Trends: A Routine Activity Approach." *American Sociological Review* 44, no. 4: 588–608.

Commonwealth v. Ellis, 708 N.E.2d 644 (Mass. 1999).

Cozens, P., and T. Love. 2015. "A Review and Current Status of Crime Prevention through Environmental Design (CPTED)." *Journal Of Planning Literature* 30, no. 4: 393–412. doi:10.1177/0885412215595440.

Danchev, S. 2016. "Was Bentham a primitive rational choice theory predecessor?" *European Journal of the History Of Economic Thought* 23, no. 2: 297–322. doi:10.1080/09672567.2014.916728.

Edelstein, A. 2016. "Rethinking Conceptual Definitions of the Criminal Career and Serial Criminality." *Trauma, Violence and Abuse* 17, no. 1: 62–71. doi:10.1177/1524838014566694.

"Evaluating Journal Articles." n.d. *Harvard Guide for Using Sources*. Retrieved February 10, 2016, from http://isites.harvard.edu/icb/icb.do?keyword=k70847andpageid=icb.page346374.

FBI.gov. 2009, September 3. "Pfizer $2.3 Billion Settlement" [Audio Podcast]. Retrieved from https://www.fbi.gov/audio-repository/news-podcasts-inside-pfizer-2.3-billion-settlement.mp3/view.

Groff, E. 2007. "Simulation for theory testing and experimentation: An example using routine activity theory and street robbery." *Journal of Quantitative Criminology* 23, no. 2: 75–103. doi:10.1007/s10940-006-9021-z.

Harrison, M. 2015. "The Terrorist's Apprentice." *Hoover Digest: Research and Opinion On Public Policy* 4: 40–45.

Istituto Per La Vigil Anza Sulle Assicurazion-IVASS. 2015. "Insurance Supervisory Authority 2015 Annual Report." Retrieved from https://www.ivass.it/pubblicazioni-e-statistiche/pubblicazioni/relazione-annuale/2016/Insurance_Supervisory_Authority_2015_Annual_Report_-_Remarks_by_the_President_Salvatore_Rossi_.pdf?language_id=3.

Insurance Bureau of Canada. 2016. Retrieved February 5, 2016, from http://www.ibc.ca/nb#.

Insurance Europe. (2016). "Insurance Data." Retrieved May 5, 2016, from http://www.insuranceeurope.eu/insurancedata.

Insurance Europe. 2016. "European Insurance—Key Facts." Retrieved from http://www.insuranceeurope.eu/sites/default/files/attachments/Europeanpercent20Insurancepercent20-percent20Keypercent20Factspercent20-percent20Augustpercent202016.pdf.

Insurance Research Council. 2011. "The Impact of Third-Party Bad-Faith Reforms on Automobile Liability Insurance Costs in West Virginia." Retrieved from http://www.insurance-research.org/sites/default/files/downloads/IRCWVBadFaith_101111.pdf.

Insurance Research Council. 2013. "New Study Finds Lower Acceptance of Insurance Fraud and Strong Support for Fraud-Fighting Efforts." Retrieved from http://www.insurancefraud.org/downloads/InsuranceResearchCouncil03-13.pdf.

Insurance Research Council. 2014. "Study Reveals the Costly Impact of Third-Party Bad-Faith Law on Florida's Automobile Insurance System." Retrieved from http://www.insuranceresearchresearch.org/sites/default/files/downloads/IRC%20Florida%20Bad%20Faith_NR.pdf.

Insurance Research Council. 2015. "Insurance Research Council Finds That Fraud and Buildup Add Up to $7.7 Billion in Excess Payments for Auto Injury Claims." Retrieved from http://www.insurancefraud.org/downloads/InsuranceResearchCouncil02-15.pdf.

International Association of Special Investigative Units. 2016. "IASIU Code of Ethics and Anti-Trust Statement." Accessed October 15, 2016, from http://www.iasiu.org/about/code-of-ethics/.

Ishida, C., W. Chang, and S. Taylor. 2016. "Moral intensity, moral awareness and ethical predispositions: The case of insurance fraud." *Journal Of Financial Services Marketing* 21, no. 1: 4–18. doi:10.1057/fsm.2015.26.

Jackson, A., K. Gilliland, and L. Veneziano. 2006. "Routine activity theory and sexual deviance among male college students." *Journal of Family Violence* 21, no. 7: 449–60. doi:10.1007/s10896-006-9040-4.

Jeffery, C. R. 1971. *Crime Prevention Through Environmental Design*. Thousand Oaks, CA: Sage.

Kula, S. 2015. "The Effectiveness of CCTV in Public Places: Fear of Crime and Perceived Safety of Citizens." *Bartin University Journal Of Faculty Of Economics and Administrative Sciences* 6, no. 12: 15–37.

Lazear, E. P. 2015. "Gary Becker's Impact on Economics and Policy." *American Economic Review* 105, no. 5: 80–84. doi:10.1257/aer.p20151107.

Leal, S., A. Vrij, G. Nahari, and S. Mann. 2016. "Please be Honest and Provide Evidence: Deterrents of Deception in an Online Insurance Fraud Context." *Applied Cognitive Psychology* 30, no. 5: 768–74.

Leukfeldt, E. R., and M. Yar. 2016. "Applying Routine Activity Theory to Cybercrime: A Theoretical and Empirical Analysis." *Deviant Behavior* 37, no. 3: 263–280. doi:10.1080/01639625.2015.1012409.

Lilly, J. R., F. T. Cullen, and R. A. Ball. 2015. *Criminological theory: Context and consequences*. 6th ed. Thousand Oaks, CA: Sage.

Luukkainen, S., K. Riala, M. Laukkanen, H. Hakko, and P. Räsänen. 2012. "Association of traumatic brain injury with criminality in adolescent psychiatric inpatients from Northern Finland." *Psychiatry Research* 200 (2/3): 767–72. doi:10.1016/j.psychres.2012.04.018.

Mckinsey and Company. 2015. "Claims Management: Taking a Determined Stand Against Insurance Fraud." Retrieved May 10, 2016, from http://www.mckinsey.com/industries/financial-services/our-insights/claims-management-taking-a-determined-stand-against-insurance-fraud.

McMahon, R., D. Pence, L. Bressler, and M. S. Bressler. 2016. "New Tactics in Fighting Financial Crimes: Moving Beyond the Fraud Triangle," *Journal Of Legal, Ethical and Regulatory Issues* 19, no. 1: 16–25.

Myers, D. G., and C. N. Dewall. 2015. *Psychology*. 11th ed. New York: Worth Publishers.

Office of the Assistant Secretary for Planning and Evaluation. 2016. "Observations on Trends in Prescription Drug Spending."

Retrieved from https://aspe.hhs.gov/sites/default/files/pdf/187586/Drugspending.pdf.

Olney, M., and S. Bonn. 2015. "An Exploratory Study of the Legal and Non-Legal Factors Associated With Exoneration for Wrongful Conviction: The Power of DNA Evidence." *Criminal Justice Policy Review* 26, no. 4: 400–420. doi:10.1177/0887403414521461.

PricewaterhouseCoopers. 2015. "The Italian Insurance Market." Retrieved from https://www.pwc.com/it/it/publications/assets/docs/italian-insurance-market.pdf.

PricewaterhouseCoopers. 2016. "Global Economic Crime Survey, 2016." Retrieved from http://www.pwc.com/gx/en/services/advisory/forensics/economic-crime-survey.html.

Ragatz, L., and W. Fremouw. 2010. "A Critical Examination of Research on the Psychological Profiles of White-Collar Criminals." *Journal Of Forensic Psychology Practice* 10, no. 5: 373–402. doi:10.1080/15228932.2010.489846.

Roden, D. M., S. R. Cox, and K. Joung Yeon. 2016. "The Fraud Triangle as a Predictor of Corporate Fraud." *Academy Of Accounting and Financial Studies Journal* 20, no. 1: 80–92.

Schneider, F., T. Brück, and D. Meierrieks. 2015. "The Economics of Counterterrorism: A Sur ey." *Journal Of Economic Surveys* 29, no. 1: 131–57. doi:10.1111/joes.12060.

Shu, L. L., N. Mazar, F. Gino, D. Ariely, and M. H. Bazerman. 2012. "Signing at the beginning makes ethics salient and decreases dishonest self-reports in comparison to signing at the end." *Proceedings of the National Academy of Sciences of the United States of America* 109, no. 38: 15197–15200. http://doi.org/10.1073/pnas.1209746109.

Skiba, J. M., and W. B. Disch. 2014. "A Phenomenological Study of the Barriers and Challenges Facing Insurance Fraud Investigators." *Journal of Insurance Regulation* 33: 87–114.

Spano, R., and S. Nagy. 2005. "Social guardianship and social isolation: An application and extension of lifestyle/routine activities theory to rural adolescents." *Rural Sociology* 70, no. 3: 414–37. doi:10.1526/0036011054831189.

Swiss Insurance Association. 2016. Retrieved May 5, 2016, from http://www.svv.ch/en/consumer-info/general-information/insurance-fraud.

Sutherland, E. H. 1949. *White Collar Crime*. Oak Brook, IL: Dryden Press.

Tennyson, S. 2008. "Moral, social, and economic dimensions of insurance claims fraud." *Social Research* 75, no. 4: 1181–1204. Retrieved from http://www.newschool.edu/centers/socres/.

Tibbetts, S. G. 2015. *Criminological Theory: The Essentials*. 2nd ed. Thousand Oaks, CA: Sage.

Tiihonen, J., M. Rautiainen, H. M. Ollila, E. Repo-Tiihonen, M. Virkkunen, A. Palotie, and T. Paunio. 2015. "Genetic background of extreme violent behavior." *Molecular Psychiatry* 20, no. 6: 786–92.

Vaske, J., D. Boisvert, and J. P. Wright. 2012. "Genetic and Environmental Contributions to the Relationship Between Violent Victimization and Criminal Behavior." *Journal Of Interpersonal Violence* 27, no. 16: 3213–35. doi:10.1177/0886260512441254.

Warchol, G., and M. Harrington. 2016. "Exploring the dynamics of South Africa's illegal abalone trade via routine activities theory." *Trends In Organized Crime* 19, no. 1: 21–41. doi:10.1007/s12117-016-9265-4.

Weber, J. 2015. "Investigating and Assessing the Quality of Employee Ethics Training Programs Among US-Based Global Organizations." *Journal Of Business Ethics* 129, no. 1: 27–42. doi:10.1007/s10551-014-2128-5.

"What is a Peer Reviewed Article?" n.d. John Jay School of Criminal Justice: Evaluating Information Sources: What Is A Peer-Reviewed Article? Retrieved March 5, 2016, from http://guides.lib.jjay.cuny.edu/c.php?g=288333andp=1922599.

Wiberg, A. 2015. "Rehabilitation of MAOA Deficient Criminals Could Lead to a Decrease in Violent Crime." *Jurimetrics: The Journal Of Law, Science and Technology* 55, no. 4: 509–26.

Xu, Z., and H. Ma. 2015. "Does Honesty Result from Moral Will or Moral Grace? Why Moral Identity Matters." *Journal Of Business Ethics* 127, no. 2: 371–84. doi:10.1007/s10551-014-2050-x.

Index

A

academic writing, 67–70
Acceptance Rate, as formula for fraud measurement, 193
ADHD (attention-deficit/hyperactivity disorder), 19–20, 33
Admiral Insurance, 176
Akers, Ronald L., 131
Albrecht, Steve, 49
Allgulander, C., 21–22
American Civil Liberties Union (ACLU), 172
analyzing measurement criteria, as step in counter fraud efforts, 121–122
Andresen, Martin, 27
anti-money laundering (AML) programs, 177
antisocial behavior, genetic and congenital foundations of, 21
antisocial personality disorder (ASPD), 20, 24
application fraud, 70, 74
Ariely, Dan, 130
assistance programs, 154–155
Association of Certified Fraud Examiners, 53, 140–141, 146, 154
attention-deficit/hyperactivity disorder (ADHD), 19–20, 33
Average Days to Investigate, as formula for fraud measurement, 194

B

baby boomers, characterizations of, 59
bad faith, as legal concept, 12–13
Bandura, Albert, 131
Beccaria, Cesare, 29, 30, 44, 184
Becker, Gary, 184
behavioral bridge
    applications, 173–177, 211
    defined, 170–172
behavioral pattern predictors (BPPs), 172–177, 211
Bentham, Jeremy, 29, 30–31, 44
big data, 109, 177, 211
billing scams, 107–108
biological school of criminology, 15, 17, 18–22, 23, 25, 32–33, 45
Bismarck System, 204
bodily injury liability coverage loss costs, 14

Boncamper, Malchus Irvin, 133
born criminal, 21
born victim, 21
BPPs (behavioral pattern predictors), 172–177, 211
brain dysfunction, and criminal behavior, 22
Bureau of Alcohol, Firearms, and Explosives (ATF), author as trainer for, 20
Business Intelligence (INFORM), 117, 118, 119
business intelligence programs, use of, 117

C

Canada, fighting fraud in, 199–200
capable guardians, 26, 27, 36, 37, 38, 115, 183
capital punishment, 185
Cartwright, A., 58
cheating behavior, honesty and, 130–143
checks and balances, 82, 141, 143–144, 145
Chen, C. X., 161, 162
city elements, 40
claims, as area to focus on when conducting vulnerability step, 80
Coalition Against Insurance Fraud, 2, 7, 128, 165
code of conduct, 144–154, 158
Code of Ethics, 151–152
cognitive flaws, as strong predictor of deviant behavior, 23
cognitive theory, 184, 185
Cohen, Albert, 42

Cohen, L. E., 26
Commonwealth v. Ellis (1999), 11
communication, as tactic in integrating controls, 116
conduct disorder, 24
control disorder, 24
Conversion Rate, as formula for fraud measurement, 193
coordination, between insurance companies and government, 11
counter fraud efforts
    developing multipronged approach to, 123–169, 211
    developing workable red flags, 66, 97–105
    identifying risks/vulnerabilities, 65–85
    integrating controls, 66, 105–116
    introduction, 64–65
    measuring, 66, 72, 75, 186–197, 211
    modifying and reassessing, 66, 116–121
    monitoring, 66, 116–121
    vulnerability assessment, 65
counterterroism financing (CTF) programs, 177
Cressey, Donald, 49, 50
crime, factors in development of, 185
crime prevention through environmental design (CPTED), 26, 39–41
Crime Prevention through Environmental Design (Ray), 40
criminal behavior, brain dysfunction and, 22
criminological theory, defined, 15

criminology, major schools of, 15, 17–31, 185
cybercrime, testing of routine activities theory in, 39

D

Dahmer, Jeffrey, 24
dangerous and unnecessary treatment, large-scale cases of, 5
Darwin, Charles, 33
dashboards, use of, 117
data
    big data, 109, 177, 211
    external data, 107, 172
    internal data, 110, 117
    mining of, 107
    as source of all effective preventative efforts, 122
determinism, 18
deterrence theory, 29–30, 44–45, 184–185, 210
deterrent effect, 36, 45, 140, 152, 184–185
developing workable red flags, as step in counter fraud efforts, 66, 97–105
differential association theory, 49
dishonesty, reduction of in virtual classroom, 138
DNA analysis, use of, 20
DNA mapping, use of within criminal justice community, 33
dopamine
    as linked to thrill-seeking behavior and attention-related problems, 22
    role of in mood and behavior, 21

DR4R, 22
DRD4, 22
Drugwatch.com, 6

E

ectomorph, 19
effective fraud strategy, capable guardians as vital part of, 37–38
ego, in Freudian approach, 23, 34
employee fraud/theft, 53, 54, 140, 141–142, 161, 162. See also internal fraud/theft
endomorph, 19
entitlement, concerns regarding younger generation's feeling of, 43–44, 59, 60, 180
entitlement theory, 62, 161–162
environmental school of criminology, 26
estradiol, as correlating to increased aggressive behavior in females, 20
Ethan Allen tour boat tragedy (2005), 132–133
ethics
    Code of Ethics, 151–152
    as priority in business dealings, 153, 154
    social learning theory and self-control theory and, 132
    triggering of internal ethics and morality, 134
ethics awareness training/ethics training, 145, 153–154, 155–156
ethics programs, 146, 155
exaggerated fraud, 73
external data, 107, 172

external threats, as area to focus on when conducting vulnerability step, 82

F

Facebook, 175, 176, 199
facial recognition systems, use of, 172
false positives, in fraud detection systems/programs, 118
FBI InfraGard, 165
Felson, M., 26
field appraisers, as area to focus on when conducting vulnerability step, 81–82
Florida, fraud legislation, 167
Forbes Magazine, 154
foreseeable danger, concept of, 41
formula of measurement, as one step in process of communicating fraud measurement, 188, 191–194
foundation of measurement, as one step in process of communicating fraud measurement, 188, 189–191
four eyes principle, 143–144
fraud
    as broader social problem, 179–181
    as category of white-collar crime, 1
    consequences of, xv
    costs of, 209
    damaging effects of, 4–6
    defined, 2–3
    as extremely difficult to quantify, xvi, 121
    lack of universal definition of, 7
    lack of universal or consistent measurement of, 71, 72, 74–75, 121
    lack of victim identification by fraudsters, 58
    measurement of, 66, 72, 75, 186–197, 211
    as number two crime in US, 1
    perception of as acceptable among medical doctors, 58
    as preferred method of financial funding for organized criminal and terrorist groups, 1, 46–47
fraud bureaus, 11, 73
fraud detection systems/programs, 106, 118, 119, 129, 205
fraud fighters, goal of, 32
fraud fighting
    application of technology in, 112
    future of, 109, 177
    opportunity reduction as key to, 61, 77
    use of data as trending wave of future of, 109
Fraud Identification Ratio, as formula for fraud measurement, 193–194
fraud policy, sample, 146–153
Fraud Savings, as formula for fraud measurement, 194
fraud triangle, 49–62, 157, 173, 210
fraudsters
    John Doe example, 125–128
    psychology of, 125
Freud, Sigmund, 23, 24, 34
fuzzy logic, 107

G

GDV (German insurance association), 203
general strain theory (GST), 42
generation X'ers, characterizations of, 59
genetics
    application of to victimization, 21–22
    influence of on crime, 20
geographic information system (GIS) data, 176
Germany, fighting fraud in, 203–206
Gilliland, K., 27
Global Fraud Study (Association of Certified Fraud Examiners, 2016), 140–141
Global Property Owners Association, 133
good faith, concept of, 12
Gottfredson, Michael, 132
Groff, Elizabeth, 27

H

hard fraud, 2–3, 73
Harrington, M., 27
Harvard University, on "Evaluating Journal Articles," 68
hiring practices, 156–159
Hirschi, Travis, 132
HIS (German insurance industry information system), 203–204
Holocaust, 33
honesty, culture of, 154–156
honesty and cheating behavior, 130–143
honesty declarations, 127, 134–137, 138–140, 143, 146, 156
honesty tests, 157
hotlines, 162–164
human component, in counter fraud efforts, 113–114
human resources, 110–111

I

id, in Freudian approach, 23, 24, 34
information sharing, as area of fraud prevention, 164–166
insurance and governmental coordination, models of, 11
insurance companies, as worst enemy, 180
Insurance Contract Act (VVG) (Switzerland), 207
insurance customers, categorization of, 128–129
insurance fraud
    costs of, 1, 2
    defined, 2
    as difficult to detect, 6–8
    as growing crime, xv, 16
    lack of research on, 130
    public acceptance of, 8–10, 37
    use of term, 73
    as very expensive to prosecute, 11
Insurance Research Council, 2
integrating controls, as step in counter fraud efforts, 66, 105–116
internal data, 110, 117
internal employees, as area to focus on when conducting vulnerability step, 82
internal ethics and morality, 134
internal fraud/theft, 54, 98, 140–144, 145, 152, 153, 156, 157, 159,

161–162, 176, 201. See also employee fraud/theft
internal training, 159–160
international perspective
    Canada, 199–200
    Germany, 203–204
    Italy, 202–203
    Malaysia, 200–201
    overview, 197–199
    Switzerland, 206–208
Internet of Things (IoT), 109, 175, 199
IQ deficiencies, as strong predictor of deviant behavior, 23
Istituto per la Vigilanza sulle Assicurazioni—IVASS, 202
Italy, fighting fraud in, 202

J

Jackson, A. K., 27
Jeffery, C. Ray, 40
John Jay College of Criminal Justice, on "What Is a Peer Reviewed Article?" 68–69

K

Krankentagegeldversicherung (daily benefits allowance), 204
Kula, S., 27

L

Leal, S., 134
legislation, as part of counter fraud policy, 166–169
Leukfeldt, E. R., 27, 39
line of business (LOB), as one step in process of communicating fraud measurement, 188
linear regression analysis, 95

link analysis, 108, 109
LinkedIn, 199
Lombroso, Cesare, 18–19
loss report process, as area to focus on when conducting vulnerability step, 80
Luukkainen, S. K., 22

M

macrogroup behavior, as focus of sociological school of criminology, 25
MADD (Mothers Against Drunk Driving), 179
Majority Model, 11
Malaysia, insurance systems in, 200–201
management reporting systems, use of, 117
mandated investigative laws, 12
mandatory auto photo inspection, legislative support needed for, 167–169
Mann, S., 134
MAOA (monoamine oxidase A) gene, 21
marketing, as area to focus on when conducting vulnerability step, 78
Massachusetts Model, 11
measurement
    analyzing criteria for, 121–122
    of fraud, 66, 72, 75, 186–197, 211
media outlets, in counter fraud efforts, 47
medical fraud, 73
medical provider, defined, 3–4
Merton, Robert, 28, 42–43

mesomorph, 19
millennials, characterizations of, 59–60
Mills, Robert, 133
Minnesota Multiphase Personality Inventory (MMPI), 24
monitoring and modifying controls, as step in counter fraud efforts, 66, 116–121
monoamine oxidase A (MAOA) gene, 21
morphology, as strong predictor of criminal behavior, 19
Mothers Against Drunk Driving (MADD), 179
Moustakas, C., 51

N

Nagy, S., 27
National Insurance Crime Bureau, 2
National Insurance Crime Bureau (NICB), 165
networking, importance of in fraud detection, 119–120
Nevada, and insurance fraud prosecutions, 12
Nilsson, B., 21–22
no-fault fraud, 3

O

occupational theft, 141–142
opportunistic fraud, 3, 42, 51, 52, 55, 73, 80, 82, 134, 198, 200, 201
opportunity
    counter fraud strategies to reduce, 41–42
    as one of three main areas of fraud triangle, 50, 60–62
    reduction of, importance of, 39, 41, 60, 61, 77, 129, 178, 200, 211
organized fraudster, 125–128

P

padding, defined, 3
peer review, 67–70
Pennsylvania, and insurance fraud prosecutions, 12
personality traits, applicability of within psychological theories of criminality, 23–24, 34
Personnel Selection Inventory (PSI), 157
Pfizer, 6
pharmaceutical industry, reduced patient care as issue in, 5–6
"A Phenomenological Study of the Challenges and Barries Facing Insurance Fraud Investigators" (Skiba), 51
planned fraud, 3, 73
pleasure versus pain, 29, 30, 31
positivism, 18
post-traumatic stress disorder (PTSD), 25
predictive analytics, 105, 106, 112
prescription drugs, spending on, 6
pressure, as one of three main areas of fraud triangle, 50, 51–54
preventative tactics, 105, 112, 120
prison reform movement, 30
prosecutorial efforts, 10–12
PSI (Personnel Selection Inventory), 157
psychoanalytic theories, 23

psychological school of criminology, 15, 17, 23–25, 34–35, 45
PTSD (post-traumatic stress disorder), 25
public awareness campaigns, in counter fraud efforts, 47
Purser, Chris, 133

Q

Quirk, James, 133

R

Räsänen, P., 22
rational choice theory, 29–31, 45–48, 143, 185, 210
rationalization, as one of three main areas of fraud triangle, 50, 54–60
red flag documents
    examples of, 98–104
    overview, 97–98
Red Flag Rate, as formula for fraud measurement, 192
red flags, developing workable ones, 66, 97–105
Redding Medical Center, federal investigation of, 5
Referral Rate (SIU Referrals), as formula for fraud measurement, 192–193
Reid, Richard, 172
Reid Report Risk Assessment, 157
Return on Investment (ROI), as formula for fraud measurement, 194
Risk Shield, 106, 118, 119
risk versus reward, 24, 29, 210
Roach, J., 58

routine activities theory, 26–27, 36–42, 48, 115, 183, 210
rule tuning, 118

S

sales, as area to focus on when conducting vulnerability step, 77–78
Sandino, T., 161, 162
Sarasota, Florida, application of CPTED strategies, 40
S.B. 318 (West Virginia, 2005) (Third Party Bad Faith Act), 13
Securities and Exchange COMMISSION (SEC), 175
self-control theory, 131, 132, 182, 183
serotonin
    as linked to thrill-seeking behavior and attention-related problems, 22
    role of in mood and behavior, 21
Shariah (Islamic Law), 201
Sheldon, William, 19
Shoreline Cruises, 133
Shu, L. L., 138
silent generation, characterizations of, 59
SIU (Special Investigative Unit), 4, 12, 61
SIU Referrals (Referral Rate), as formula for fraud measurement, 192–193
SIU-fraud units, as area to focus on when conducting vulnerability step, 80–81
social environmental theory, 41
social learning theory, 131–132
social media, 79, 175, 176, 177, 199

social network detection, 107
social network visualization, 108
sociological school of criminology, 15, 17, 25–31, 32, 35–48, 209–210
soft fraud, 3, 73
software detection world, 106–107
somatyping, 19
Spano, R., 27
Special Investigative Unit (SIU), 4, 12, 61
statistical analyses, 95
strain theory, 27–28, 42–44, 173, 210
street-robbery setting, testing of routine activities theory in, 38–39
subrogation (recovery) units, as area to focus on when conducting vulnerability step, 81
superego, in Freudian approach, 34
Sutherland, Edwin, 49, 50
Sutherland, Edwin H., 131
Switzerland, fighting fraud in, 206–208

T

Takaful Act of 1984, 201
takaful fund, 201
technology, as cost-effective preventative tactic, 105, 111, 112
terrorists, characterization of by the public, 46
testosterone, as correlating to increased aggressive behavior in males, 20
text mining, 107
theories
    cognitive theory, 184, 185
    deterrence theory, 29–30, 44–45, 184–185, 210
    differential association theory, 49
    entitlement theory, 62, 161–162
    fraud application of, 32
    general strain theory (GST), 42
    rational choice theory, 29–31, 45–48, 143, 185, 210
    routine activities theory, 26–27, 36–42, 48, 115, 183, 210
    self-control theory, 131, 132, 182, 183
    social environmental theory, 41
    social learning theory, 131–132
    on social structure and anomie, 28
    strain theory, 27–28, 42–44, 173, 210
    vulnerability theory, 181–183
Third Party Bad Faith Act (West Virginia S.B. 218, 2005), 13
transparency, 154, 161, 162
traumatic brain injury (TBI), and criminality, 22
Twitter, 175, 199

U

Underwriting, as area of focus on when conducting vulnerability step, 79
underwriting fraud, 74
up-coding, 4, 86, 88, 89, 100

V

vacations, as time when internal fraud cases are discovered, 145
vendors and outsourced partners, as area to focus on when

conducting vulnerability
step, 81
Veneziano, L., 27
Veracity Analysis Questionnaire
(VAQ), 157
victim blaming, 21
victim facilitation, 21
victim precipitation, 21
victimization
genetic factors in, 21
link of to opportunity rather than
premeditation, 41
vulnerability assessment
assessing vulnerabilities, 75–85
completing document for, 85–97
as step in counter fraud efforts, 65
Vulnerability Assessment Worksheet,
75, 76, 85–96
vulnerability theory, application of,
181–183

W

wages, and internal theft, 161–162
Warchol, G., 27
white-collar crime
cognitive level of criminals in,
46, 47
as considered victimless
crimes, 58
defined, 4
fraud as category of, 1
psychological profile of, 124–130
Works Constitution Act of 1952
(Germany), 204

Y

Yar, M., 27
Yates, Andrea PIA, 24

Yearly Operating Expense, as formula
for fraud measurement, 194

Z

zero tolerance, 45, 57, 78, 129, 140, 143

CPSIA information can be obtained
at www.ICGtesting.com
Printed in the USA
FFHW021831201118
49518094-53878FF